Matter and Form

Matter and Form

From Natural Science to Political Philosophy

Ann Ward

LEXINGTON BOOKS

A division of

ROWMAN & LITTLEFIELD PUBLISHERS, INC.
Lanham • Boulder • New York • Toronto • Plymouth, UK

Published by Lexington Books
A division of Rowman & Littlefield Publishers, Inc.
A wholly owned subsidiary of The Rowman & Littlefield Publishing Group, Inc.
4501 Forbes Boulevard, Suite 200, Lanham, Maryland 20706
http://www.lexingtonbooks.com

Estover Road,
Plymouth PL6 7PY
United Kingdom

British Library Cataloguing in Publication Information Available

Library of Congress Cataloging-in-Publication Data

Ward, Ann.
Matter and form : from natural science to political philosophy / Ann Ward.
p. cm.
Includes bibliographical references and index.
ISBN 978-0-7391-3568-6 (cloth : alk. paper) — ISBN 978-0-7391-3570-9 (electronic)
1. Political science—Philosophy. 2. Natural history—Philosophy. 3. Philosophy of nature. 4. Nature and civilization. I. Title.
JA71.W28 2009
320.01—dc22 2009027476

Printed in the United States of America

∞™ The paper used in this publication meets the minimum requirements of American National Standard for Information Sciences—Permanence of Paper for Printed Library Materials, ANSI/NISO Z39.48-1992.

For Mary
My sea of joy

Table of Contents

Acknowledgments

The present volume arose from the workshop that I chaired for the 11th International Conference of the International Society for the Study of European Ideas (ISSEI), *Language and the Scientific Imagination*, held at the University of Helsinki in Helsinki, Finland, 28 July–2 August 2008. I wish to thank Dr. Marja Harmanmaa of the University of Helsinki for organizing and convening the conference and the ISSEI for sponsoring it. I would like to give special thanks to Ezra Talmor and Rachel Ben-David of the ISSEI for inviting me to develop and chair the workshop that has given this volume its name. The ISSEI has provided me on more than one occasion the opportunity to pursue my research in collaboration with others such that I would not be the scholar I am today without the society. I also wish to express my gratitude to the participants of the workshop and the contributors to this volume. It is their expertise and far-reaching intellectual interests that have made this volume possible. I have benefited both professionally and personally from their research and hopefully their friendship.

I also wish to acknowledge the University of Regina President's Fund for awarding me a Social Sciences and Humanities Research Council (SSHRC) General Research Grant, and Campion College at the University of Regina for awarding me two Campion College President's Research Awards, all of which provided generous financial support for my participation in the 11th International Conference of the ISSEI and for preparation of the manuscript of this volume. I also wish to express my gratitude to my research assistants, Kris Schmaltz, graduate of the MA Program for Social and Political Thought at the University of Regina, and Even Dechief, student in Political Science at the University of Regina. In the beginning stages of the project Kris provided assistance in researching and presenting the general concept of the volume, and in the latter stages of the project Evan provided much needed help in the composition of the index. The diligence, technological skill, and intellectual curiosity of both Kris and Evan were invaluable in bringing this project to fruition.

Finally, I wish to thank my husband and colleague Lee Ward, to whom I owe all things, and to welcome our new daughter Mary into our world.

Regina, June 2009

—Ann Ward

Introduction

Ann Ward

In Plato's *Phaedo* we learn from Socrates that while previous philosophers who inquired into material nature sought to explain the motion of matter, they could not explain the cause of matter. Thus, while Presocratic natural science explored how things do what they do, it abstracted from the question of why things do what they do and why such action is best. The insufficiency of Presocratic natural science led Socrates to focus on the soul and to articulate the theory of Ideas or Forms as an account of the ultimate source of causality in the cosmos. Aristotle further developed Socrates' teleological understanding of nature, explaining the motion of matter through his discussion of the four causes and his concepts of form and natural place. Moreover, Aristotle's teleological natural philosophy has a profound impact on his political philosophy. Aristotle argues that human beings are by nature political animals, having their natural end or place in the city whose regime is hierarchically structured based on differences in moral and intellectual capacity. Medieval political theorists from the monotheistic faith traditions attempted to synthesize classical natural and political philosophy with the revealed truths of sacred scripture. They argued that divine reason structures an ordered universe the awareness of which allows for psychic and political harmony among human beings.

Early modern and Enlightenment thinkers challenged the natural philosophy of classical and medieval philosophers, ushering in a more liberal and egalitarian political and moral order. For instance, Bacon argued for the conquest of nature to relieve the human estate. Cartesian skepticism and scientific methodology required the autonomy of the individual thinker. A clear example of how this new science led to a new politics is Hobbes. He argued that bodies, once in motion, are eternally in motion until hit by other bodies that are themselves in motion. Thus, there is no rest in nature as there are no Aristotelian forms or natural places that govern matter. Hobbes applies his mechanistic understanding of material nature to his understanding of human nature. Individuals in the state of nature are naturally free and equal and driven by their passions into an endless pursuit of one power after another until death. However, from this mechanistic understanding of humanity's natural condition, Hobbes develops a social contract theory in which civil and political society is constituted from consent. Later thinkers such as Locke and Rousseau modify this Hobbesian premise in their pursuit of the protection of rights and a free society.

Materialist and mechanistic conceptions of the cosmos, however, have not always given rise to liberal democratic social and political philosophies. The influence of historicist ways of thinking on scientific inquiry in the 19th century, for example, can be connected to Charles Darwin's theory of evolution. Reflecting on the struggle for existence among all organisms in the natural world, Darwin reasoned that there is a process of natural selection that over time facilitates the production of ever newer and better adapted species as it causes the extinction of others. Influenced by what can be called a form of social Darwinism, Friedrich Nietzsche explores what he regards as the diminishment of the human species brought about by the democratic movement. For Nietzsche, the democratic movement, having its origins in Socratic philosophy and Christian morality, must be confronted by philosophers of the future who, by replacing the moral categories of good and evil with the amoral concepts of strength and weakness, call forth a new type of nobility. Nietzsche envisions an aristocratic order that draws its inspiration from art rather than the rationalism embodied in the history of natural and political philosophy. The chapters in this volume address these and related themes as they explore the relationship between natural science and political philosophy from the classical to contemporary periods.

Although continuing and building upon the recent scholarship that considers the impact of scientific ideas concerning the natural, material world on the ethical and political ideas of philosophers in the history of philosophy, this volume also adds to it in significant ways. For instance in *A Philosophy of Political Myth* (2007), Chiara Bottici provides an insightful analysis of the history of scientific thought in Classical Greece and early modern and Enlightenment Europe. Bottici argues that classical political philosophy did not have a word to designate the concept of "physical reality," and that for Plato being is found in the ideas that are both thoughts and the objects of thought. Therefore, Plato articulates an ideal realm of greater reality than the visible, material world of which the ideal realm is the intelligible cause. Along with the devaluing of physical reality, classical political philosophy, according to Bottici, does not distinguish between myth (*mythos*) and reason (*logos*). Thus, in the writings of Plato and Aristotle no association is made between myth and untrue speech.

Turning to early modern and Enlightenment Europe, Bottici argues that Descartes, one of the founding fathers of the Enlightenment's scientific revolution, in contrast to Plato, reduces the concept of "idea" to simply a thought; it is something of "mental content" lacking any certain objective reality. This reduction follows the separation of the knowing subject from the object of knowledge, or the thinker from the external world. Moreover, modern science proposes that the external world can only be made intelligible to the thinking subject if it is conceived mechanistically, as simply matter in motion. The intelligibility of this external, material world can only be plumbed by the subject if reduced to mathematical language and laid bare to sensory experience by empirical demonstrations. Anything that cannot be reduced to mathematical figures and proved by experiment is merely subjective and therefore unreal. According to Bottici, since myth cannot be reduced to numbers or tested empirically, it is also said to be "unreal."

Bottici argues that the thoroughgoing materialism of the Enlightenment concept of reality was motivated by the attempt of science to liberate itself from the authority of Christianity. Yet, it also coincided with the struggle for human autonomy, identified by the Enlightenment with "pure reason." However, as with classical political philosophy's postulation of an ideal, immaterial realm, Bottici argues that the Enlightenment's belief in the purity of reason also turns into a mythology of its own. Moreover, Bottici argues that other early modern and Enlightenment thinkers such as Hobbes, Locke, and Rousseau had self-conscious recourse to "myth" when they postulated their various images of the "state of nature."

The most significant way in which this volume differs from Bottici's is in its stance toward the place of the mythical in the history of philosophy. In essence, Bottici claims that the history of scientific and political thought is peopled not by philosophers but by mythmakers, or rather that the dichotomy between the truthful and the mythical must and did break down. Alternatively, this volume argues and rests on the premise that in both the natural and political philosophies of classical antiquity, early modernity, and the Enlightenment, the distinction between the rational and the mythical, true and untrue discourse, is clearly maintained. Indeed, many of the contributors to this volume, including David Schaefer, Paul Ulrich, Leah Bradshaw, Ingrid Makus, Lee Ward, and Steven Robinson, suggest that early modern, Enlightenment, and contemporary thinkers believe that their rational reflections on the natural, material world prevents the mythical and irrational from slipping into their reflections on the ethical and political world. For instance, Ulrich and Bradshaw argue that Hobbes sought to illustrate that he was correct about the "natural condition" by explicitly making proofs based on natural science and by offering a conception of knowledge that is grounded in empirically verifiable determinants.

In our contributions, moreover, Douglas al-Maini, Coleen Zoller, Mostafa Younesie, Ingrid Makus and I, suggest that although classical antiquity may not, as Bottici argues, have a had a word to designate the concept of "physical reality," classical political philosophy did indeed have a rich understanding of and interest in this reality. While acknowledging that for classical political philosophers matter is ordered teleologically toward the "ideas" or "Forms" or toward a transcendental notion of the Good, these contributors also recognize that rational comprehension through mind of the structure, order and processes of the material world is for these philosophers both possible and may even be necessary if humanity is to achieve its highest end.

In *Leaving Us to Wonder: An Essay on the Questions Science Can't Ask* (2005), Linda Weiner and Ramsey Eric Ramsey examine the influence of scientific thought on the human world. They question the common assumption among scientists that the scientific method is the only way to proceed toward answering important questions, and most importantly, that science is the only legitimate foundation for ethics. Weiner and Ramsey focus on Western philosophical thought from the Enlightenment through to its critics, analyzing the relation between the material and the ethical, as well as truth and action, with the objective of achieving a greater understanding of truth and the good life. Like Weiner and Ramsey this volume suggests an alternative to extreme positions that privilege science or philosophy over one another. However, it does not, as Weiner and Ramsey do, provide a normative critique of scientific thinking. Rather, this volume explores the relationship between scientific and ethical and political thought in the history of philosophy. Moreover, it reaches beyond the Enlightenment and its critics, to include chapters that reflect on natural science in the ancient and medieval worlds.

Friedel Weinert, in *The Scientist as Philosopher: Philosophical Consequences of Great Scientific Discoveries* (2004), deals directly with the philosophical consequences of scientific discoveries, specifically with discoveries in physics. Weinert portrays the interaction between science and philosophy as dialectic. Scientists approach problems with metaphysical presuppositions having to do with nature, physical understanding, time, causation and determinism, which are then tested. The result either confirms or refutes such presuppositions. Philosophers must then decide whether to alter metaphysical conceptions. Weinert gives specific attention to philosopher scientists: those research scientists who engage in philosophy by participating in discussions about the consequences of their discoveries. Alternatively, this volume focuses not on the philosophical work of scientists, but rather on the scientific work of philosophers. Moreover, whereas Weinert emphasizes how scientific discoveries change our perception of the material world, this volume places emphasis on how changing scientific ideas affect our perceptions of morality and politics.

In *Consilience: The Unity of Knowledge* (1998), Edward O. Wilson advocates induction from different evidential bases as a method that can produce agreement among all branches of learning, and thus unify the sciences and the humanities. Wilson calls this state of unity consilience. The book is organized around the premise that the Enlightenment tradition begun by Francis Bacon of the intrinsic unity of knowledge, coupled with the belief in human progress, needs to be reinvigorated. Wilson suggests that intellectual unity will be advanced if the method of reductionism used in the sciences is used in other forms of knowledge. This method of study, according to Wilson, will lead toward a large-scale consilience in understanding of all things as the result of universal laws of physics.

Unlike Wilson, who reduces complex human behaviors and functions such as artistic creation and morality to the outcome of genes governed by physical laws, this volume includes chapters on ancient and medieval thinkers whose natural philosophies do not lead to a mechanistic reductionism in their moral and political philosophies. For these philosophers physical order in the material world is consistent with and even supports an independent psychic harmony in the spiritual world. This volume also includes chapters that illustrate the divergence between the scientific ideas and the ethical and political theories of early modern, Enlightenment, 19th century and contemporary philosophers. It also contains chapters that question the plausibility of the arguments of postmedieval philosophers such as Hobbes who specifically try to ground their political prescriptions in a mechanistic understanding of the physical universe.

Recently, many good books have addressed the significance of the revolutionary thought of Francis Bacon (1561–1626). *Francis Bacon and the Refiguring of Early Modern Thought,* edited by Solomon and Gimelli (2005), explores how Bacon sought to unify knowledge diachronically, across historical time, and synchronically, across social and intellectual space. With respect to unifying knowledge across time, the volume considers Bacon's integration of earlier forms of thought into his new philosophic system. For instance, it investigates the role that the ancient tradition of atomism, founded by Democritus, played in grounding Bacon's ratio-empirical philosophy, and the extent to which Bacon borrowed the concept of "*eidos*" or "Form" from ancient Neoplatonism and transformed it into his notion of a materially embedded "Law" of nature. The volume also explores Bacon's attempt to link the various scientific disciplines across social and intellectual space by investigating his belief in a primitive philosophy of matter or an inductive method that undergirds all knowledge, thereby unifying the sciences and the humanities. However, *Francis Bacon and the Refiguring of Early Modern Thought* does not focus on the relationship between Bacon's natural philosophy and his political ideas.

Robert K. Faulkner and Julian Martin make the interconnectedness of Bacon's natural and political philosophies central to their books. In *Francis Bacon and the Project of Progress* (1993), Faulkner argues that Bacon's project of a new experimental science does not integrate but rather refutes the scientific theories of the ancients. Moreover, the new science is definitively tied to a more practical aim, the foundation of a progressive civilization peopled by self-made men who assert their power over nature and the universe. The progressive and scientific civilization that engages in an ever-expanding conquest of nature will be grounded in a modern state characterized by a limited, republican government that pursues empire abroad while ensuring private security and economic and industrial growth at home. Martin, in *Francis Bacon, the State, and the Reform of Natural Philosophy* (1992), argues, like Faulkner, that Bacon's natural philosophy cannot be separated from his political goals. Bacon's program for the reform of the natural sciences was subordinate to and put in the service of his program for reform of the state. However, the reformed modern state, the conditions for which made possible by the new science, was not republican in character, but rather was conducive to a centralizing, imperial monarchy.

Although Faulkner and Martin recognize the crucial connection between natural philosophy and political philosophy, their focus is solely on the thought of Francis Bacon. This volume is much broader in scope. It reaches back to the natural and political philosophies of ancient and medieval thinkers, including the Presocratics, Plato, Aristotle, Ptolemy, and Averroes, for new insights. It also looks forward to other early modern and Enlightenment philosophers and their critics, such as Montaigne, Descartes, Hobbes, Locke, Rousseau, and Nietzsche, as well as contemporary theorists of morality and politics.

What is Europe? European Democratic Culture, edited by Rieu and Duprat (1995), studies the rise of European democratic thought. It takes account of how the political emerged in ancient Greece, the Copernican Revolution and the conditions that led to modern political thought, further developments such as natural rights and problems of sovereignty, and contemporary developments in social and political philosophy. It argues that democracy arises out of specific historical conditions that include revolutions in thought that change the conceptions individuals have of themselves in relation to the world and society. These revolutions in thought are a result of scientific changes that arise out of certain social and cultural changes. By scientific revolutions Rieu and Duprat mean an intellectual discipline, the source of which is the European scientific tradition. Because scientific discoveries are the product of certain social changes, science and philosophy exist in interrelation. Democracy is thus envisioned as grounded in historical conditions rather than *a priori* principles.

Like *What is Europe?,* this volume explores the relationship between science and political thought in the history of the West, with the exception of Ahmed Abdel Meguid's chapter that explores this relationship in the thought of the Islamic philosopher and theologian Averroes. This volume, however, is not focused narrowly on democratic thought only, but investigates how science informs the ethical and political thought of philosophers more broadly. Moreover, it does not argue for the historicist basis of changes in scientific and political philosophy.

In *Nietzsche and Modern Times: a Study of Bacon, Descartes, and Nietzsche* (1993), Laurence Lampert explores what he calls the two "great events" in the West's "spiritual drama": the rise of modern science made possible by the revolutionary theorizing of Bacon and Descartes, and the philosophy of Nietzsche, the first philosophy, according to Lampert, to reveal a comprehensive understanding of the modern era initiated by Bacon and Descartes. Although beginning with Bacon and Descartes, the book moves toward and finds its main interest in Nietzschean philosophy which, for Lampert, is postmoral in the sense that it is situated in and takes its guidance from the natural history of the universe brought to light by science. Lampert argues that Nietzsche's postmoral direction is necessitated by the fact that Bacon and Descartes radically altered the setting of philosophy by making science a public discipline. All previous philosophers eschewed such publicity as contrary to the flourishing of human communities that require the fictions of permanence and exclusivity. Denying what he believes is a deeply ingrained misconception that Nietzsche was an enemy of science and an advocate of art, Lampert argues that Nietzsche advances a science that breaks with both the Platonic science of the transcendence of nature and the Baconian science of the mastery of nature. Rather, Nietzschean philosophy embraces what Lampert calls a science of "pure immanentism or naturalism" that is wholly consistent with the naturalistic worldview of contemporary evolutionary biology. Nietzsche, according to Lampert, calls for the embrace of the earth as the natural home of transient beings.

Lampert's study is a profound analysis of the relationship between Nietzsche's philosophy and his understanding of the natural history of the universe revealed by science. This volume, however, suggests that the embrace of the contemporary worldview of evolutionary biology need not be as total as Lampert implies. Rather, Steven Robinson seeks a middle ground for the naturalization of political theory between the truths derived from evolutionary biology and those derived from theoretical reflection on the history of human morality and politics. Thus, Robinson does not reject in its entirety the history of political philosophy, and draws classical and early modern political theory to explain human behavior. Moreover, in an alternative reading to Lampert's, this volume includes chapters by David Schaefer, Dwayne Raymond, Juhana Lemetti, and Leonard Sorenson, who argue that not only Nietzsche but other early modern and Enlightenment philosophers such as Montaigne, Descartes, Hobbes, and Rousseau can be understood as incorporating the insights of ancient philosophy.

Karen Cordrick Haely, in *Objectivity in the Feminist Philosophy of Science* (2008), analyzes prominent feminist critiques of objectivity in science. These feminist critiques, Haely argues, have exposed the sexist and/or andocentric biases in science. In order to enrich the concept and practice of objectivity in light of the current biases that diminish it, Haely proposes that a properly constituted scientific community be composed of a diversity of individuals who can bring the perspective of marginalized persons into view. A central theme of Haely's book, therefore, is that political and ideological commitments, such as those that stem from gender and racial identities, influence scientific theories and practices. Unlike Haely's *Objectivity in the Feminist Philosophy of Science,* the contemporary philosophy of science is not the main subject of this volume, and issues of gender and race, although related themes, will not be organizing themes. However, like Haely's book, it does explore the relationship between politics and science.

This volume takes a unique interdisciplinary and international approach with contributions from scholars in the fields of Philosophy and Political Science. The contributors teach and research in places as diverse as Canada, the United States, Europe, and the Middle East. Part one explores ancient conceptions of natural science in its relation to ethics and politics. Doug Al-Maini, in chapter 1, explores the social role of Presocratic natural science in polis-era Greece. Presocratic natural philosophers begin the ongoing analysis of what later came to be known as "matter" and "form," a type of analysis Al-Maini refers to "hylomorphism." Al-Maini's thesis is that Presocratic hylomorphism "naturalizes" the polis, or gives theoretical grounding and justification to the discursive nature of Greek politics. Focusing on the work of Anaximander and Xenophanes, of particular interest to Al-Maini is the idea that forms—natural elements—find their home in matter—Anaximander's *apeiron* or Xenophanes' primordial earth-mixture. Al-Maini argues that the model of choice when describing this aspect of hylomorphism is the polis itself, the polis providing a home for citizenship and the plurality of political interests, as matter provides a home to forms. According to Al-Maini, Presocratic natural philosophers, products of the polis in their thinking, projected a reflection of the polis onto nature. The polis, in turn, became for the Greeks the instantiation in the social world of nature's character. Presocratic natural science, therefore, serves the purpose of helping to stabilize the polis and its political forum as a legitimate form of social organization.

In chapter 2, I explore Socrates' use in the *Phaedo* of Presocratic natural science to show the immortality of the soul. The argument for the immortality of the soul developing out of Presocratic natural science occurs within Socrates' first argument from contraries. I argue that the problem with the first argument from contraries, as presented by Socrates, is that although the inquiry into material nature shows *how* things do what they do, it does not show *why* things do what they do. Specifically, it does not show why the soul, and all other living things, first came into being, thus initiating their motion. The first argument from contraries thus serves to illustrate the fact that human questions, such as whether or not the soul is immortal, turn into questions of natural science, such as why and how the cosmos came into being. Socrates' second argument from contraries resolves the problem from Presocratic natural science through reference to the Ideas or forms. The Ideas as cause in the second argument from contraries are inseparable from their particular manifestations in the natural world. Yet, I argue that the Ideas so understood cannot demonstrate the immortality of the individual soul, but only the cosmos of which the soul is a part. I conclude by arguing, much as Al-Maini does for the Presocratics, that Socrates' Ideas ground not only the cosmos but also the *polis.* By making the discursive practice of politics possible the Ideas show how the soul flourishes in this life if not the next.

Coleen Zoller, in chapter 3, explores the role of mathematical-scientific understanding in living well in Plato's philosophy. Zoller argues that not only does Plato think that "doing science" improves our understanding of the cosmos; he also believes that looking at life with an eye toward science improves our understanding of ourselves in relation to others. For instance, in *Republic* II, Socrates shows that an individual *polis* and the global networks of political associations can only flourish in an environment where citizens use the art of measurement to calculate the impact of their lifestyle on both the health of oneself as well as the rest of the world's citizens. Yet, the healthy city turns feverish when the appetite for luxury develops, the pursuit of which leads to war. In the *Gorgias,* Socrates envisions that the end of war will come once the citizens of the world adopt what Zoller calls Plato's "science of living well." The science of living well requires physical and psychic training that teaches citizens to measure what they really need, seeing beyond mere appearances of what they want. Zoller argues that in both dialogues we see why failure to master the art of measurement leads to both physical and psychic disease as well as political conflict. Taking a fresh look at Plato's mathematical-scientific grounds for living well will enhance our efforts today to flourish as individuals and citizens of the world.

In chapter 4 Mostafa Younesie argues that Aristotle's understanding of politics as a science is demonstrated by reflecting on the order of the books in the *Politics* in light of Aristotle's *Physics.* Specifically, if we consider the different books of the *Politics* as various discussions of "matter" and "form," terms and concepts introduced in book 2 of the *Physics,* then a rational ordering of these books into a homogeneous unit known as the *Politics* becomes evident. Focusing on books 1 to 3 and 7 to 8 of the *Politics* as examples, Younesie argues that books 1 and 2 present the polis in its universal meaning as the human entity par excellence. In this "special" human entity Aristotle gives us human beings as householders and citizens, both of whom prove to be the "matter" of the polis. Being simply human is not sufficient and, if the polis is to achieve its highest end, the human beings within it must be transformed into citizens. Citizenship, however, is defined by the political structure or constitution, discussed in book 3, which governs the polis. The best or normative form of citizenship, according to Younesie, is defined by the best constitution. Thus, human beings and citizens, as the "matter" of the polis, must adopt the constitution, as the "form" of the polis, to effect the transformation to normative citizenship. This movement from human being to citizen to normative citizen is imitated in the *Politics* in the movement from books 1 to 2 and then to book 3. However, book 3 elaborates various constitutions and thus different types of citizenship without pinpointing the best constitution and hence the normative type of citizen. In books 7 and 8, on the other hand, Aristotle reflects on the polity composed of the best citizens living the best way of life (book 7) under the best constitution (book 8). Yet, Younesie argues that it is education, not the constitution, which comes to light in book 8 as the actual agent or "form" that transforms the citizen as such into the ideal citizen of the ideal polity.

The Ptolemaic and medieval Islamic understandings of the ordering of the soul in relation to the cosmic whole or divine being is explored in part two. In chapter 5, Michael Weinman analyzes Ptolemy's claim in the *Syntaxis* that there is a causal connection between the mathematical study of the heavenly appearances and a well-ordered soul. Ptolemy's claim is of special interest because it emphasizes that only as one undertakes the study of the heavens as a mathematician, and not as a theologian or a physicist, can one so improve the state of one's soul. Weinman investigates what exactly about the mathematical study of the heavenly motion underwrites this claim. Focusing on

Ptolemy's account of the uniform if anomalous motion of the sun, Weinman finds that mathematical understanding means precisely *not* looking for the path (or orbit) of the sun, but rather seeing the uniformity in the unchanging character of the motion as a whole, not as an unfolding movement from here to there and back again, but as one unchanging thing: the very being of order. It is in coming to understand this about the sun and its motion that Ptolemy means for his work to contribute to the ordering of our souls as students of mathematics. Weinman concludes that the heart of Ptolemy's project—standing as it does on an anti-mechanistic, but also nontheological, life-world—is lost on us today in our having divided accounts of the universe into "scientific" and "religious," with no possibility of bridging the two realms.

In chapter 6 Ahmed Abdel Meguid argues that in his commentary on Plato's *Republic,* Averroës asserts that the ultimate foundation of political science should be psychology. Against this background he sets out to eliminate what he saw as Plato's dialectic and so unscientific claims in the *Republic.* Meguid thus suggests that the ground for Averroës' critique of what he regarded as Plato's claim about *necessity* in politics, particularly regarding the trajectory of change from one political constitution to the other, is based on Averroës' psychology. Building on Alfred Ivry's and Muhammad 'Abid al-Jābrī's claims that Averroës wrote both the *Middle Commentary* on the *De Anima* and the *Commentary* or *Summary* of Plato's *Republic* toward the end of his career, Meguid bases his claims about Averroës' psychology on the *Middle* rather than the *Long Commentary* on the *De Anima.* In this vein the chapter argues that Averroës' stance on the material intellect in the *Middle Commentary* underlies his criticism of Plato's claim that his layout of the five possible constitutions and the relation between them is necessary. Meguid concludes by indicating that his reading of Averroës' critique reflects how Averroës held a more liberal political outlook in comparison with the conservatism that tended to pervade medieval political philosophy.

The theme of part three is the early modern reconception of nature, both human and nonhuman, as mechanistic, and how this new understanding transformed the foundations of legitimate moral and political order. In chapter 7 David Schaefer explores Michel de Montaigne's philosophic skepticism. Although having affinities to the classical political philosophy of Plato and Aristotle rather than to the Pyrrohnians with whom Montaigne is commonly associated, Schaefer argues that Montaigne's skepticism is distinctively modern in that it is linked to his espousal of a new doctrine of nature that came to underlie the modern scientific quest. Adopting a deterministic and materialistic account of nature, Montaigne's skepticism emphasizes the necessity of grounding our lives on our perceptions and indicates the possibility of constructing a science of medicine on the basis of systematic observation of nature. Montaigne's emphasis on the analysis of sense data as the ground of a progressive natural science is consistent with if not the cause of his hedonist moral and political teachings. He promotes a popularized public skepticism toward the rightness of religious opinions with the aim of turning most people away from the quest for transcendence and making them "at home" in the terrestrial world, in which the only real ills to be avoided are physical pain, disease, and premature death. For philosophers, Montaigne, according to Schaefer, endorses a libertine way of life that eschews all restrictions on sensual pleasure except where the liberty of the other is concerned.

The Harvey-Descartes debate over the motion of the heart turns on their divergent explanations for the hardening of the heart. Whereas Harvey maintains that it hardens while contracting, Descartes maintains that it hardens while expanding. In chapter 8 Dwayne Raymond addresses recent scholarship that argues that the debate between Harvey and Descartes is largely metaphysical and not methodological. Specifically, he analyzes G. Gorham's recent argument that the debate centers on the different views that Descartes and Harvey held with respect to the mind-body relation. Raymond argues that while much is correct in Gorham's view, its strict dichotomy fails to appreciate that, for Descartes, method and metaphysics are intimately linked. According to Raymond, the division between mind and body and the laws of motion, specifically Law B, which Descartes employs in his detailed explanation of the heart's motion, result from his unique development of a set of logical intuitions drawn from Parmenides' method, Aristotle's logic, and Euclid's *Elements.* Raymond argues that Descartes' unique development of these logical intuitions provide the structure for his tightly reasoned arguments. Moreover, this development is related to his conception of the proper use of the will. As such, it explains Descartes' conjecture that yeast may lie in the folds of the heart and his unwillingness to endorse Harvey's view of a self-contracting heart despite devastating observation: the heart continues to expand and contract once removed from the body.

Paul Ulrich, in chapter 9, analyzes Hobbes' description of the "natural condition" of mankind in *Leviathan.* Ulrich argues that although Hobbes makes clear that the natural condition is a condition of war, this does mean that it is a condition of ongoing conflict but rather a condition in which nothing prevents conflict while a known disposition to conflict exists. An explanation for the origin of the disposition to conflict can be found, according to Ulrich, in Hobbes' teaching on the materialistic, mechanistic workings of the human mind. His early chapters on perception, sense, memory, thought, and speech seek to prove that all thought begins with the actual physical contact of an external body on an organ of sense. Thus, the mind appears as a unique system dependent on the functioning of one's organs and on one's unique set of experiences, and consisting of one's particular set of passions. Ulrich concludes that wisdom, for Hobbes, is the sense each individual has of the coherence of the totality of their experiences and perceptions. Competition and therefore the disposition to conflict originates in the physical impossibility of conceiving of wisdom superior to one's own, which naturally appears to oneself as complete or perfect wisdom.

In chapter 10, Leah Bradshaw draws on Hobbes and Aristotle as archetypes of two ways of knowing, with a view to understanding how these ways of knowing impact on the crafting of political institutions. Hobbes begins from the universal human experience of sensation, rooted in the sensory organs' collision with external, material forces. From this epistemological basis, Hobbes, according to Bradshaw, argues that he can demonstrate unequivocally that he has a foundation for politics that is scientific and universally true for all people in all time. For Hobbes, autonomous, desire-maximizing individuals must trade absolute natural liberty for the peace of the state. Building on Hannah Arendt's analysis of his political philosophy, Bradshaw argues that Hobbes is a remarkably *unpolitical* thinker. Encouraging an increasingly privatized citizen absorbed in his own personal fate, Hobbes' simultaneously constructs a commonwealth that is imperialistic and homogenizing, its unlimited expansion of power destroying any political particularity in the international system. Unlike Hobbes, Aristotle, Bradshaw argues, believes that all human action and thought inclines toward purpose. For Aristotle, to know anything is to inquire, through reason, into the purpose of its existence, or its *telos*. Aristotle, according to Bradshaw, is thus a much more *political* thinker than Hobbes, as his philosophy of purpose grounds considerable variability in political associations. Bradshaw concludes that the universalism of Hobbes' "one world" politics rests on a world science that is based on a questionable foundation. Aristotle's conception of *telos,* on the other hand, offers us a more expansive way of conceiving of political variability in a world of difference.

Juhana Lemetti, in chapter 11, suggests that Hobbes' reaction to Aristotle is not absolutely antithetical. Rather, Lemetti stresses that Hobbes' primary target was not so much Aristotle as the late scholastic Aristotelianism that attempted to adapt Aristotelian philosophy to Catholic dogma. Nonetheless, Lemetti illuminates various points of disagreement that point to Hobbes' attempt, in contrast to Aristotle, to articulate a pluralistic view of human nature and its purposes. The divergence between Aristotle and Hobbes centers on Hobbes' critique of what he considers Aristotle's false metaphysics. Unlike Aristotle's positing of two substances in nature, matter and form, Hobbes posits only one, matter. For Hobbes, there are no Aristotelian forms or essences in nature. Thus, the core of the disagreement is Hobbes' rejection of Aristotle's essentialism. The critique of essentialism, Lemetti argues, plays out on four levels in Hobbes' philosophy: psychology, ethics, politics, and human happiness.

In chapter 12, Ingrid Makus argues that an Hobbesian "science" of politics that challenges Aristotelian assumptions is flawed. Attempting to establish familial relations on the basis of artificial covenant, Hobbes cannot adequately explain the perpetuation of the human species and the reproduction of the political community. Although Aristotle's emphasis on the natural biological basis of familial relations at the origin of the polis would seem to exempt him from such difficulties, Makus argues that Aristotle's account suffers from problems as well. Hobbes shows us what is missing in Aristotle—an acknowledgment of the deliberative or reasoning capacity of women—whereas Aristotle shows us what is missing in Hobbes—a notion of maternal love and care for one's own. Recognizing the limitations of both the Hobbesian and Aristotelian models, Makus concludes, can inform our discussion of contemporary issues surrounding the politics of biology and reproduction.

The scientific roots of modern liberalism and contemporary social Darwinism is the theme of part four. In chapter 13, Lee Ward examines an important, but often neglected, aspect of Locke's liberal philosophy. On the one hand, in Locke's political writings the moral principle of natural equality appears to rely on the fact that human beings all belong to the same species. Yet in his more philosophic *Essay Concerning Human Understanding,* Locke systematically dismantled the concept of "species," reducing it to an arbitrary linguistic category. Ward argues that the attack on the traditional Aristotelian notion of substance as a foundational principle of intelligibility was in service of Locke's defense of a new conception of natural philosophy based on experiment, hypothesis, and probability. This, Ward argues, creates a problematic parallel between Locke's natural and moral philosophy. While Locke claims that moral principles are capable of demonstrative knowledge, unlike the more provisional findings of natural science, Ward maintains that ultimately Locke's effort to narrow the range of what can be considered knowledge means that both natural and moral philosophy operate in the cognitive realm of probability, as opposed to knowledge strictly speaking. Ward concludes by reflecting upon the implications for liberal society derived from Locke's understanding of the relation between natural and moral philosophy. He speculates whether Locke's humbler and more analytical form of philosophy is capable of maintaining the uneasy balance between natural and moral philosophy operating on distinct ideas of truth. This "divided mind" of liberal society contains the possibility of giving rise to factions moral and religious, on the one hand, and scientific and secular, on the other, doing battle over the very meaning of what constitutes evidence, knowledge, and proof.

Leonard Sorenson, in chapter 14, argues that an exploration of Rousseau's botanical works reveals his philosophical critique of modern natural science. Rousseau believes that something important about nature, including human nature, can be learned from the study of plants. Unlike reductionist modern science, Rousseau argues that plants are heterogeneous wholes with parts and, like all embodied beings, have structured motion toward certain ends: self-reproduction and death. Moreover, for Rousseau the proper way to study plants exhibits the proper way to study simply, and the study of the beings, such as plants, constitutes the actualizing of the natural potentials of the soul and mind of the human being. Sorenson argues that in his *Discourse on Inequality Among Men,* Rousseau, as in his botanical works, casts his account in the language of natural potentials, arguing that the human soul and mind, like plants, have the capacity for internally driven "self-perfection" and externally influenced "perfectibility." Thus, as some plants are the best expression of their kind, the exemplar of humanity is he or she who advances furthest in self-knowledge of oneself in relation to the whole and other human beings. In his botanical works Rousseau, according to Sorenson, exhibits himself at the peak of human knowledge as his study of plants teaches him the law of nature of all embodied beings and makes him self-aware of his place in the life-cycle. Sorenson concludes that in botany we see the emergence of the first and highest purpose of political philosophy: the quest for knowledge.

In chapter 15, Steven Robinson explores a fundamental problem for all of us in the modern West who are committed to democracy. Although we adhere to the natural goodness of some version of democratic and egalitarian political morality, we cannot help but acknowledge that humans throughout history have been a notoriously undemocratic species. Robinson looks to both the biological and the humanistic approaches to human nature to shed light on this problem. The biological approach is reflected in traditional social Darwinism and contemporary neo-Darwinian theory. Both schools of thought tend to see human nature as essentially selfish and view democratic egalitarianism as an evolutionary "mistake" or temporary maladaptation within human biology. This dark view of human nature and socio-political possibilities is confirmed by the contemporary science of comparative primatology. In contrast to the biological approach, Robinson argues that the humanistic approach understands human beings as psychologically complex, torn by competing tendencies toward selfish authoritarianism and altruistic egalitarianism. Robinson explores three early modern thinkers, Niccolò Machiavelli, Etienne de la Boetie, and William Shakespeare, who reflect this human complexity. According to Robinson it is through the contemporary social science of cultural anthropology that the competing biological and humanistic accounts of human nature can be brought together. Cultural anthropology argues that through the development of complex language prehistoric human communities maintained an egalitarian political ethic. The development of language, in short, allowed for the articulation of an ethic of cooperative opposition to dominance by high-status individuals. Thus, Robinson concludes that the slow and painstaking development of anti-authoritarian and democratic values within recent history can be understood as a reassertion of the prehistoric egalitarian ethic.

The chapters in this volume engage in a timely exploration of the relationship between natural science and ethical and political philosophy. The thought of not just one philosopher or philosophic period is studied, but the thought of individual philosophers and related philosophies in the history of philosophy is considered, from the classical to contemporary eras. They thus illuminate the importance of philosophic reflection on the natural, material world to moral and political theorizing, and show how these two ways of thinking are interrelated in the thought of the philosophers studied. The volume thus seeks to mediate between the "two cultures" of the sciences and the humanities, making a contribution to ending the isolation that separates them.

BIBLIOGRAPHY

Bottici, Chiara. *A Philosophy of Political Myth.* Cambridge: Cambridge University Press, 2007.

Faulkner, Robert K. *Francis Bacon and the Project of Progress.* Lanham, Md.: Rowman & Littlefield, 1993.

Cordrick Haely, Karen. *Objectivity in the Feminist Philosophy of Science.* New York: Continuum International Publishing Group, 2008.

Lampert, Laurence. *Nietzsche and Modern Times: a Study of Bacon, Descartes, and Nietzsche.* New Haven: Yale University Press, 1993.

Martin, Julian. *Francis Bacon, the State, and the Reform of Natural Philosophy.* New York: Cambridge University Press, 1992.

Rieu, Alain-Marc, and Gerard Duprat, eds. *What is Europe? European Democratic Culture.* London; New York: Routledge, 1995.

Solomon, Julie Robin, and Catherine Gimelli Martin, eds. *Francis Bacon and the Refiguring of Early Modern Thought.* Aldershot, UK: Ashgate Publishing, 2005.

Weiner, Linda, and Ramsey Eric Ramsey. *Leaving Us to Wonder: An Essay on the Questions Science Can't Ask.* Albany: State University of New York Press, 2005.

Weinert, Friedel. *The Scientist as Philosopher: Philosophical Consequences of Great Scientific Discoveries.* Berlin; New York: Springer, 2004.

Wilson, Edward O. *Consilience: The Unity of Knowledge.* New York: Vintage Books,

Part I

ANCIENT SCIENCE, NATURAL TELEOLOGY, AND THE ORDER OF POLITICS

Chapter 1

The Polis Philosophers

Douglas Al-Maini

At the end of an often neglected fragment, Xenophanes laments the recognition he receives from the state, especially when compared to the honours heaped on athletes victorious at the Olympic games: "and even if he were to win with horses he would get all these, not being as worthy of them as I. For our expertise is better than the strength of men and horses. But this practice makes no sense, nor is it right to prefer strength to this good expertise" (Fragment 2).[1] Xenophanes' conclusion rests on the observation that athletic victories do not markedly contribute to the benefit of the polis, while his own wisdom does.[2] Somewhat jarringly, he even goes on to name the field in which his wisdom better serves the polis than does athletic prowess: politics. The fragment continues, "Neither if there were a good boxer among the people, nor if there were a pentathlete or wrestler . . . would for this reason a city be better governed." Xenophanes thus raises the question of how he in particular, and by extension possibly all philosophers, have utility to the government of the polis. This paper is devoted to answering the question implicit in Xenophanes's bold assertion.

Based on reports of service, one answer to the question relies on the most direct route possible: philosophers use their wisdom to craft good laws and policies. Strabo, Plutarch, and Diogenes Laertius indicate that Parmenides served as a legislator for Elea,[3] and this seems to be corroborated by archeological discoveries in the region.[4] Similarly Thales is thought to have had some kind of role in governance.[5] But this can be only a part of an answer to the riddle posed by the fragment of Xenophanes. For if the great service provided by these philosophers lies in the laws they wrote, we might wonder why, rather than a record of those laws, history has left us a preponderance of their fragments concerned with cosmology and cosmogony. Indeed, the preservation of such esoteric work itself begs to be explained, considering the abstract and impractical nature of the texts. Especially when compared to the usefulness and applicability of constitutions, early speculation over the primary materials of the physical universe seems to be of trivial importance to the flourishing polis. Surely the record of interest in what we might call metaphysics and physics argues against understanding the earliest philosophers as important because of a legislative role.

Furthermore, this preliminary interpretation cannot help us to explain the relevance of Xenophanes himself or of other philosophers that are not legislators. Xenophanes is the prototypical wandering philosopher and, not being a citizen in any of the cities he visited, would be unable to directly participate in a foreign city's political process. As he himself tells us, "already there are seven and sixty years, tossing about my counsel throughout the land of Greece," and "I tossed about, bearing myself from city to city" (Fragments 8 & 45). To take the skeptic's case further still, what are we to make of the philosophers that positively avoided public service? Heraclitus seems to dismay over the prospects of common people ever understanding his work.[6] He disavows any interest in politics, and yet his philosophy is seized upon and becomes famous throughout polis-era Greece.

In trying to understand the social role of philosophy,[7] Jaeger championed the view that these early thinkers facilitated a cultural transition to a more analytical and rational perspective.[8] This answer gives the philosophers the Herculean task of changing the mindset of the polis from irrational to rational, and the ability of any such small minority being able to effect such a transition is questionable. Recent work has revealed the foundational importance of debate to the polis from the very beginning, signaling a deep, structural disposition towards good argumentation.[9] Thus it is not the case that early philosophers effect the transition themselves, but rather that they participate in the unfolding of the inherently discursive nature of the polis. Furthermore, Jaeger's methodological response and others like it ignore the actual content of Presocratic philosophy. What is desired for a fuller understanding of the matter is an interpretation that can explain philosophy by referring to what it actually asserts.[10]

The immediate problem faced by anyone looking to delve into the place of philosophy within Greek society lies in the Presocratic fascination with nature. A predilection for cosmology seems a long way off from being relevant to that most human and artificial of constructs, the polis. However, the polis presents its own unique set of demands for being understood, especially when we turn to its discursive nature. The need for a theoretical grounding of Greek politics is a pressing issue in a way not possible in authoritarian contexts. One identifying marker of polis culture is the extent to which it directs itself at the question of which constitution is best, and the basic options range from full-fledged democracy through various broadly based and limited oligarchies down to the tyranny of one. Such a debate is unheard of in this period outside of the argumentative confines of the polis: in well established contemporaneous authoritarian cultures the rule of many does not warrant serious consideration. What makes the debate different within the polis is the real possibility of a change in political structure, should the results of the debate warrant it. The history of Athenian tinkering with its own constitutional structure bears this out, as does the sheer variety of constitutions present within the polis system. That being said, the reality of the polis as a forum within which meaningful discourse can take place presents real limits to this constitutional debate. Simply put, should the polis decide to embrace tyrannical rule, it would no longer be a polis. There is an inherent tension, then, between the need to take seriously the various options available within the argument concerning which constitution is best, and the need to preserve the debating forum as an arena for relatively balanced interests competing with each other. For the forum itself to persevere, it needs persuasive grounds for itself as an interactive process between competing but balanced concerns.

It is into this gap that the early philosophers and their cosmological interests can be fitted. Presocratic philosophy begins the ongoing analysis of what would come to be called matter and form. Henceforth, this particular brand of analysis will be referred to as "hylomorphism"; the term is a combination of the Greek word for matter, *hyle,* with the word for form, *morphe.* In scholarly literature "hylomorphism" is most commonly used to refer to Aristotle's work, but no such particularized reference should be understood here. It is the thesis of this paper that the hylomorphism of Greek philosophy provides the basis of an argument justifying the order of human society peculiar to the Greeks. Thus the debate between philosophers over fundamental physics reprises the constitutional debate of the polis. At stake is an argument that shows either harmony or dissonance between polis and universe, providing powerful theoretical ammunition in the question of the constitutional form the polis should adopt. Armed with a view of nature that shows it to be a context in which various elements compete but never overwhelm each other, Greeks could look to the polis as the instantiation in the social world of nature's character. The import of this endeavour should not be understated: our hypothesis is that hylomorphism naturalizes the polis. The specific goal of this paper is to outline how the earliest philosophers of Greece accomplished this goal. In particular, we shall look at the work of Anaximander and Xenophanes.

Since the concepts of form and matter have such a rich history within the polis period, before proceeding to outline the case for our hypothesis it is necessary to state what we mean by these terms. "Matter" and "substance" are catch-all words used to identify the thing in which forms are found. Briefly, "form" refers to the determinate characteristic that language can describe, and "matter" is a term for the thing that can take on a variety of forms. Another way to understand form and matter makes use of the concepts of actual and potential. Since matter can take on a variety of forms, matter is sometimes described as being "pure potentiality": it is potentially anything, and is nothing determinate in its own right. A form, on the other hand, is determinate. It is as if all the potentiality of matter were minimized and we were left with a single, definite quality or set of qualities. Of course, the concept of hylomorphism is fraught with theoretical difficulties, and, as mentioned, the variations in Greek philosophy on the basic theme are many. Of particular interest to this paper is the relationship between form and matter and the idea that forms find a home in matter. Given the political perspective adopted in this paper, the model of choice when describing this aspect of hylomorphism is of the polis itself. There is a broad variety of looks that a polis can make manifest in the world, and each polis is constantly changing its look. Thus while the actual determinate make up of a polis like Athens, for example, may change and develop through time, we have no difficulty in continuing to call that changing thing Athens. And yet, in a prosaic sense, Athens is just the determinate qualities and characteristics we can point to and describe. This dual nature is at the heart of hylomorphism. On this view, Athens is both the matter or substance in which all these changes occur and the various definable sets of characteristics that we refer to when we speak of Athens.

Bearing this analysis in mind, we may proceed to the earliest philosophers. The first conceptions of the universe look decidedly hierarchical, and there is a temptation to conclude a dissonance between them and the principles underlying the polis. Thales famously posits water as a "first principle," Anaximander puts forward a mysterious concept called the *apeiron,* and Anaximenes champions air as the prime element. It is interesting that these models occur at the very earliest phase of natural philosophy, and that we know the least about them. It is therefore unclear just how hierarchical these natural systems are, and caution is required in assessing their place in this analysis of the early cosmologies. However, the system of Anaximander is markedly different from that of the other two in hypothesizing a plurality of elements making up a fluid structure that will not collapse into uniformity.[11] Thus the famous quote from Simplicius: "for they [the elements] give penalty (*dike*) and recompense to one another for their injustice (*adikia*) in accordance with the order of time—speaking of them in rather poetical terms. It is clear that, having noticed how the four elements change into one another, he decided not to make any of them the underlying thing, but something else beside them."[12] In explaining

Anaximander's reasoning, Aristotle tells us that if one of the elements were primary, then it would destroy the others, or at least that the others would have been completely converted into it by now (recognizing, as Anaximander does, that the elements do change into each other). But of course we now observe all the different elements, and therefore no one of them is primary.[13]

With this move the understanding of the universe as a changing array of constituents comes to the fore. In Anaximander's cosmos, the basic elements oscillate between growth and remission. Elements change, for example water becoming air and air eventually changing back into water. As mentioned, in his physics Anaximander posits that such a fluid system requires a stable basis in something else, and this he names the *apeiron,* or "the boundless." It is tempting to think of the *apeiron* as being a primary element, and that Anaximander is in fact positing a hierarchical physics, but to do so would be to ignore what Aristotle states to be the reason for speculating about the *apeiron* at all: Anaximander cannot accept that any one of the perceivable elements expressing the process of change could be more primal or powerful than the others. We might wonder why this necessary condition on the nature of the elements calls for the supposition of something so radically different, the *apeiron,* and why Anaximander's cosmos cannot consist solely of elements changing into each other without some mysterious "boundless" entity undergirding the change. In other words, what prevents us from positing a constellation of elements, none of which are primary, but which do directly change into each other, as opposed to a route via the *apeiron?* Of course that process of change would need to be explained, but such a model would seem to eliminate the need for the *apeiron,* since there is still no dominating element to worry about. Anaximander is concerned with particular elements going out of existence such that eventually the universe settles down into being one kind of stuff, since this outcome clearly goes against what we observe. If, as just hypothesized, elements could directly change into each other, it is at least conceivable that an element might disappear as a result of something like neglect, i.e., at some particular moment all of it changes into other elements but none of the other elements happen to be in the process of turning into it. For fear of this outcome one might feel justified in abandoning our hypothesized universe. But surely it would be odd in such a system for none of the remaining elements to turn into the missing element ever again, given that they had done so in the past. So even if an element did seemingly disappear for a period, there would be nothing to prevent its recurrence. In this way the worry over devolution into a monistic physics would be averted, and the question remains: why conjecture the *apeiron?*

The dearth of evidence makes any answer to this question necessarily speculative, but the description of the *apeiron* in the various fragments is suggestive. As mentioned, the literal translation of the word *apeiron* is "without bounds" or "without limits," and we can interpret this in at least two ways. First, we can think of the boundless as something like an empty set or as having no content at all. Surely the concept of "limit" requires that there be something which has the limit, and so one way of understanding "without limit" is to assert that there is in fact nothing being referred to, and hence an absence of limit as well. We have good reason to think that this is not how Anaximander intended his concept of the *apeiron* to be interpreted, for we are told that the *apeiron* is the origin of all existing things and "also what existing things die back into."[14] Our second, preferred option then, is to consider the *apeiron* as having content, but this again leaves us with several interpretive options. First, it could be limitless in the sense of being without end. Like a field that one could never find the limits of in order to put a fence around, the *apeiron* is unbounded. Second, we can think of the concept of "limit" as a tool by which we are able to distinguish and mark differences. According to this perspective, the *apeiron* cannot be described by concepts that function to define objects by separating out the characteristics they have from the characteristics they do not have. Because it takes in *all* characteristics the *apeiron* itself would be indescribable, as it could not be compared to or differentiated from anything else. On the other hand, it could be considered as the totality of all the elements, something like the whole of the cosmos. On this view, we should not think of the *apeiron* as indistinct, a sort of homogenized mass of indeterminateness, but rather as the totality of being which has heterogeneous and changing constituents.[15]

Interpretive questions such as these bedevil Anaximander scholars, but they need not concern us overmuch. What matters here is the assertion of an elemental content to the *apeiron.* That being the case, Anaximander's fluid process of change amongst the elements begins to look very much like the rough outline of hylomorphism given above. The *apeiron* serves as a substrate to change, and none of the elements are actually destroyed in the most radical sense of that term, i.e., going right out of existence. Rather, they remain in the *apeiron* as potential. At any particular moment, the universe we see has an arrangement of elements in various stages of preponderance and diminution, but this is subject to change and so a host of other arrangements are there in potential. Indeed it seems as though the whole notion of elements "changing into each other" in Anaximander's philosophy is misleading. Rather, what we experience as the elements are, in Anaximander's words, "separated out" from the fullness of the *apeiron.* This generation of elements does not invoke the creation of something out of nothing, a worry that will become a dominant theme in the philosophy of a future Presocratic, Parmenides. Neither is any element ultimately destroyed, and this helps makes sense of the concern raised by Aristotle. If the model we proposed in place of Anaximander's were actually to be the case, then the problem of generation and destruction becomes acute, as does the demand to understand the process of how one element changes into another. That model really does have to contend with the problem of something coming from nothing. Anaximander insures that no element goes out of existence because the *apeiron* allows for change without actually entailing destruction; rather, an element returns to be a part of the *apeiron.* In this way the concern with devolution is met in as strong a manner as possible.

Seen in this light, the role of Anaximander's *apeiron* looks like that of the polis providing a home for citizenship. The competition between the physical elements is akin to the competition between citizens trying to persuade each other in the decision-making process of the polis. After all, persuasion is nothing more than the attempt to change others in their beliefs, and the same citizen may have an argumentative history of swinging back and forth between various positions. In Anaximander's system elements change back and forth, but can only do so by means of having the *apeiron* in place to support that change. A polis also manifests a similar fluidity in the changing allegiances that operate between citizens, factions, and interests in its own peculiar history of politics. However, all of the argumentative positions and the various arrangements and factions that citizens take on are there in potential once we recognize the polis as the setting of debate. As an institution of persuasion the polis requires difference among its constituents. In this way the surface plurality of interests we see operating in deliberation requires a polis in place to undergird it, just as Anaximander's cosmos of fluctuating elements requires the *apeiron.* Without a difference of opinion, no debate occurs, and even if some policy receives unanimous backing, to sustain such unanimity would entail the practical end to the polis as a discursive forum. Or rather, the unanimity possible in a polis is of a different type than that achieved in authoritarian regimes, since a polis acts from consensus. Dissonance and factionalism are far more prevalent than unanimity in the polis, but this is not to say that any particular polis cannot be unified despite such differences. Realizing that the next dispute brings with it new opportunities and the possibility of success can be a motivation to accept defeat; in other words citizens on the losing end of a debate might think to themselves that they are currently in a state of remission, but that their influence may grow in the future. So too for Anaximander the elements "pay each other back according to the order of time."

This account of behaviour on the part of the elements has obvious antecedents in consensual politics. Interestingly enough, the dissent essential to the polis is submerged by the end of the decision-making process. However consensus is achieved (such as the Spartan practice of determining who the loudest shouters are in its Assembly), the remarkable fact about a successful polis is its ability to come to a policy decision and achieve unity out of difference. Thus a well-functioning polis will always have some members that must accept policy decisions going against their own beliefs, wishes, and limited interests. The key here is the legitimate status the polis (and the peculiar type of political process it makes manifest) has in the minds of its citizens. If members can accept polis-practice as a legitimate decision-making process, then the polis will be unified. Rejection of the practice of consensus undermines the entire system in a way more perilous than seeming dissent does in authoritarian regimes. The polis depends on the candid participation of its members because in the polis citizens are vested with power and must be persuaded; in authoritarian systems dissent can be eliminated by force, and the threat of this makes dissent and real political difference all but absent within a hierarchy. In the polis citizens must be persuaded to ratify the political system itself with the full understanding that they may not have their own views becoming policy. This last condition helps explain why any argument that grounds the polis and persuades citizens that the polis is a legitimate form of government would be of special value. The similarities between Anaximander's cosmology and the structure of the polis need to be seen in this light, as Anaximander provides an understanding of the universe that implies the naturalness of the polis.

Xenophanes also engages in physical speculation, but unfortunately his theory is not well preserved and any account of it will necessarily involve a significant degree of reconstruction. However, a few observations can be made with some confidence. Xenophanes seems to base his system on only two elements, earth and water; everything else consists of some state of one of these elements or their mixture.[16] Thus Fragment 29 states that "All things which come into being and grow are earth and water," and Fragment 33, "For we all come into being from earth and water." With these two elements in place, Xenophanes has the means to explain all the physical phenomena he experiences, as he proceeds to do, giving an analysis of clouds, rain, rivers, rainbows, and even wind; all these phenomena are explained by reference to their elemental base of earth and water. Thus clouds and wind are formed out of the sea, rain falls from the clouds, rivers form and flow back into the sea, the source of water (Fragment 30). Similarly, "And she whom they call Iris [i.e., the rainbow], this too is by nature a cloud, purple, red, and greenish-yellow to behold" (Fragment 32).

So far so good, but how are we to square this elemental dualism with the fragment that states, "For all things are from the earth and to the earth all things come in the end" (Fragment 27)? This fragment seems to posit only one element. Indeed, the ancient testimonia for Xenophanes comfortably describes Xenophanes as a monist, especially in his religious thinking, and this argues against the picture being presented here. More primary evidence is needed to clear up decisively our interpretive tension, but on the issue of his natural philosophy, one attempt to make the various fragments coherent proceeds by accusing Xenophanes of equivocating over the term "earth." To understand this response, it will be helpful to look at another fragment plus an ancient commentary. First, Fragment 28: "This upper limit of the earth is seen here at our feet, pushing up against the air, but that below goes on without limit (*apeiron*)." Most interesting for our purposes, being now somewhat familiar with Anaximander, is the reference to something *apeiron,* and its identification with earth. Second, part of the report on Xenophanes from Hippolytus:

Xenophanes thinks that a mixture of the land with the sea comes about, but that in time the land becomes free from the moisture, and he asserts that there are proofs for these ideas: that shells are found inland and in mountains, and he says that in quarries in Syracuse imprints of fish and seals were found; and in Paros the imprint of coral in the deep of the marble and on Malta slabs of rock containing all sorts of sea creatures. He says that these things came about when long ago everything was covered with mud and then the imprint dried in the clay. And he says that all men will perish when the land sinks into the sea and becomes mud, at which time generation begins again, and this change comes about in all worlds.[17]

The equivocation seems to be that when in Fragment 27 Xenophanes speaks of "earth" he really refers to a mixture of the elements water and earth,[18] but in Fragments 29 and 33 he refers to the actual element earth. This kind of equivocation is not uncommon and something like it occurs when we think of a cloud as gas; technically a cloud is a mixture of gaseous atmosphere with water droplets. In fact it seems like the same problem in Presocratic philosophy occurs with another word meant to designate an element, "air." "*Aer*" originally referred to both atmosphere and mist, and eventually came to be narrowed to meaning only atmosphere.[19] Similarly, sand or dust is earth with its water content dried off, but we still refer to the obviously damp soil of agricultural land as "earth." Separating these two senses of the word earth (element and combination of elements) allows us to take Fragments 27, 18, and the testimony of Hippolytus together to generate a cosmogony. For Xenophanes, a mixture, what Hippolytus refers to as "mud," is made up of both earth and water. The elements are then separated out from this soup, although they will return to it again, as Hippolytus asserts, and this will ensure the end of all men and indeed anything whose being is predicated on the separation of the elements. At that point the process begins again.

How much of a hylomorphic theory this is has been questioned, the key issue being whether or not we need to reserve the mixture of Fragment 27 as a special substance, different in kind from other mixtures. The alternative to this kind of full-blown hylomorphic theory is the option of the mixture-earth being something like what we moderns, with our atomic view of the world consider a mixture to be. On this view, mixture is juxtaposition, wherein elements are not mixed in any fundamental way, but rather are placed in varying degrees of closeness to each other and in various proportions. Of course it would be anachronistic to interpret Xenophanes's physics as a form of atomism. Instead he may have just been making the observation that all physical stuff literally returns to the earth at some time (but does not undergo any fundamental change in being combined with the other element). This view is sometimes referred to as a "geophysical" interpretation of Fragment 27.[20] To some extent deciding how much of a hylomorphic theory Xenophanes has depends on how one reads the reference in Fragment 28 to the *apeiron.* Those willing to speculate over the influence of Anaximander here would be amenable to understanding Xenophanes's mixture-earth as akin to Anaximander's *apeiron,* and so a mixture of a more radical and fundamental sort.[21] Since Xenophanes does still call it earth and gives it a physical location, one is left supposing that some amount of this primordial stuff could be picked up, but that we would be unable to distinguish the earth from the water within it. And indeed, there may be some experiential basis for this interpretation, since some examples of mud can present an extremely homogenized appearance.[22]

Supporters of the geophysical model would point to the fact of Xenophanes's usual practice of describing only the most ordinary of phenomena is not in keeping with an abstract analysis of an underlying fundamental matter. It takes a leap of interpretive faith to move from an observation that the sun passes over the earth and spreads warmth over its surface (Fragment 31), for example, to the heady metaphysical speculation we are involved in. Furthermore, there is nothing in the fragments of Xenophanes to indicate he believed that the elements changed into each other (as seems to be the case with Anaximander). However, the other fragments mentioned so far indicate that Xenophanes was interested in providing the most profound understanding of nature that he could, and the passage from Hippolytus outlines a mixture of earth and water that underlies all change that occurs. For Xenophanes, the mixture is the state from which all distinguishable characteristics arise, and it is the destination of all objects as well. Key to the issue, it seems, is whether any object or indeed any element is able to survive the periodic creation and destruction of the world.[23] If nothing survives and the world is entirely separated out from the mixture, as Hippolytus seems to indicate, then Xenophanes's mixture amounts to being a kind of matter anyways, since the phenomenal world is effaced in the mixture and the mixture is potentially any of the various forms it will take on as differences in structure arise. Taken to its logical conclusion, this interpretation would entail that even the elements themselves would not be distinguishable within the mixture; rather they are there in potential, waiting to be separated out as it were. It is tempting to think that on this view Xenophanes is expressing the belief that not only does the earth stretch away indefinitely beneath our feet in Fragment 28, but also that it is *apeiron,* that is, indefinite in the sense that we are unable to distinguish elemental differences within it.

Be that as it may, it is clear that Xenophanes's analysis of observable phenomena is based on a theory of mixture, and that the earth provides a matrix from which different structured complexes of the elements arise. Thus Xenophanes is at least minimally a hylomorphist, in the sense that he is willing to talk about some basic material that undergirds all change. Furthermore, his plurality of elements, once separated out from the matrix of earth, are engaged in a continual process of combination and separation with each other. For the same reasons as were given in the case with Anaximander, we can fit Xenophanes into our rubric of providing a background cosmology that naturalizes the polis. Because both systems rely on a foundational principle, the *apeiron* for Anaximander, the primordial earth for Xenophanes, it is tempting to see these systems as hierarchical in nature, setting one element aside as different and somehow superior to the others. However, this is misleading, and, relying on the analogy that is being made here between city and cosmos, we can say this would be akin to mistaking the city for a citizen. A basic category mistake is made in these judgments since the *apeiron* in Anaximander and the earth-mixture in Xenophanes are a different kind of substance than the elements that combine to produce the experienced world. Indeed in both cases these foundational principles contain the elements in their most integrated and, one might even say, egalitarian form. The recognizable interplay between citizens has been abstracted and transformed into the interactions of the physical elements, even to the point of different elements becoming preponderant under different circumstances within the phenomenal world, just as different associations of citizens become preponderant within the polis. At a second step of abstraction, an utterly idealized citizenry can be recognized in the underlying *apeiron* and earth-mixture. Here nothing dominates, all is mixed to so great an extent that the elements are not readily distinguishable, and we are left with a sense of stability and equality. The *apeiron* and the earth-mixture sit in the background of the play of elements, supporting the continuous change that occurs between the elements. Similarly, the real polis has a legislative history displaying the rough interplay between citizens, but it is predicated on the idea of the polis, which requires an expression of difference in its populace and so must insure the possibility of that difference in its own being.

Here we can give but a brief survey of the other philosophers, but it will suffice to provide a view of the general trend. Empedocles gives an account of the four cardinal elements (earth, fire, air, water) plus the forces of love and strife. These elements and forces are continuously at play: "in this respect they come to be and have no constant life; but insofar as they never cease from constantly interchanging, in this respect they are always unchanged in a cycle" (Fragment 25).[24] It is true that Heraclitus prioritizes fire, but that is because fire is change par excellence; Heraclitus will not allow for static elements of any kind within his system. He does, however, stress the importance of a balance that is created between opposing forces: "what opposes unites: the finest attunement stems from things bearing in opposite directions, and all things come about by strife," (Fragment 8) and "they do not understand how, while being at variance, it is in agreement with itself. There is a back-turning connection, like that of a bow or a lyre" (Fragment 51).[25] Aristotle gives us a list of ten opposites that the Pythagoreans believed were somehow constitutive of the world, beginning with limited/unlimited. Apparently it was the interaction of this pair that creates the reality of the elemental numbers constituting the world.[26] The Atomists provide perhaps the most radically pluralistic universe of them all, going so far as to say that "the eternal existents consist in minute substances, infinite in number."[27] Most distressing is the apparent loss of concern for the whole in this system: a full account of the atoms is enough to explain the universe, and the void, within which atoms combine and interact, is given the troubling status of "nonexistent." Furthermore, "no plurality, Leucippus said, could arise from what is truly multiple."[28] It seems that for the Atomists, the problem of the one and the many is solved by disavowing any ontological status to the one. Even Parmenides, the great monist of the early philosophical tradition, finds it fitting to combine his "Way of Truth" with a much longer "Way of Seeming," in which a cosmology of elements presents itself: "But since all things have been named light and night, and these have been applied according to their powers to these things and to those, all is full of light and obscure night together" (Fragment 9).[29] In describing the universe that we live in, Parmenides fits squarely within the polis-validating tradition and, close to the beginning of the Way of Truth, even has this to say about the Way of Seeming: "But nevertheless you shall learn these things as well, how the things which seem had to have genuine existence, permeating all things completely" (Fragment 1). While these thinkers express great variety in their thought, the uniformity of this elemental aspect to their natural philosophies deserves emphasis. Returning to our original fragment, we can broaden the discussion to conclude that the early philosophers and their wisdom are a boon to government because they show just how legitimate traditional polis governance is; by extension, they highlight the need to protect and preserve the fundamentals of the system. Indeed, it is for this reason that they deserve to be called the Polis Philosophers, as their work serves the purpose of helping to stabilize the polis as an established form of social organization.

So far we have explained early philosophy as providing the theoretical groundwork of the polis, but of course, the process is operating in the other direction as well. In other words, it is no surprise that philosophers who are products of the polis in their thinking should find a reflection of the polis in nature; they are far from providing an unbiased and "scientific" examination of the universe.[30] If, as is being argued here, the project of natural philosophy serves the interest (conscious or not) of providing the Greek social world with a persuasive justification for its own makeup, then the criteria for acceptable results on the part of the polis philosophers are already well established. But we should not be too hard on this projection of political realities onto nature, for while it does circumscribe how the cosmos is to be understood, it also liberates the Greek mind from the more usual Mediterranean ways of thinking. The polis philosophers tell us that, when dealing with the interaction between arts and sciences, as often as not there is an underlying harmony of perspective that drives both worlds. Hence the advent of natural inquiry goes hand in hand with the rise of a culture of rhetoric and persuasion, and indeed supports that culture by validating its ground rules.

ENDNOTES

1. All translations and fragment numberings from J. H. Lesher, ed., *Xenophanes of Colophon: Fragments: A Text and Translation with a Commentary* (Toronto: University of Toronto Press, 1992). For a concordance with other numbering systems of the fragments, see Lesher, *Xenophanes*, 10. Lesher calls Fragment 2 a "famous comment" on the glory given to Olympic victors, and this is fair, given the attention it has received in scholarly literature as a remarkable example of a Greek thinker questioning the value of athletic festivals. That being said, the fragment fails to appear in most modern collections of the earliest Greek philosophers. Thus Robin Waterfield, ed., *The First Philosophers: The Presocratics and the Sophists* (New York: Oxford University Press, Oxford World's Classics Edition, 2000); Patricia Curd, ed., *A Presocratics Reader* (Indianapolis: Hackett Publishing Company, 1996); etc. The choice to leave out Fragment 2 (and Fragment 1, for that matter) shows the continuing influence of the interpretation of these early philosophers as physicists unconnected to nor concerned with the polis, and so their social observations are pushed into the background; on this interpretive stance.

2. The best analysis of this unusual attack on athletics in general and the Olympics in particular is still C. M. Bowra, "Xenophanes and the Olympic Games," *The American Journal of Philology* 59, no. 3 (1938), 258–79.

3. Strabo, VI 1.1; Plutarch, *Reply to Colotes* 1126a; Diogenes Laertius (hereafter DL), 9.23.

4. A. D. Trendall, "Archeological Reports for 1966–67." *Journal of Hellenic Studies* 87 (1967): 32–33.

5. Herodotus I 75.4–5 gives him a role in establishing the structure of Ionian government.

6. Although on this point we run into the problem of accepting Diogenes's biography. Diogenes mentions nothing about public service, and it is difficult to see what ulterior motive of Diogenes the anecdote serves. Heraclitus' low opinion of the intelligence of common people is a theme he returns to again and again; see T. M. Robinson, ed., *Heraclitus: Fragments* (Toronto: University of Toronto Press, 1987), fragments 1, 2, 5, 15, 17, 19, 29, and 34; for his reticence over serving in office, see D.L. 9.3. Heraclitus's obvious elitism complicates the matter of his willingness to be of any use to the polis, instead possibly wanting his work to remain in the private hands of a select few. Regardless, his work and thought enters the public domain and becomes part of the intellectual background to the polis. As such, it is worthwhile questioning if and how the thought of Heraclitus is of use to the polis.

7. As Lesher puts it, "which aspects of Xenophanes's teachings would have been especially conducive to good government?" (Lesher, *Xenophanes*, 57).

8. So Werner Jaeger, *Paideia: The Ideals of Greek Culture* Trans. G. Highet, 3 vols. (Oxford: Oxford University Press, 1939–45) 173.

9. Most importantly, see Jean-Pierre Vernant, *The Origins of Greek Thought* Trans. J. Lloyd (Ithaca, 1982).

10. For a summary of recent attempts at explaining Xenophanes's specific intentions with fragment 2, see Lesher, *Xenophanes*, 57.

11. Vernant, *Origins*, 119–27, gives a fuller analysis of Anaximander along these lines.

12. Simplicius, *Commentary on Aristotle's Physics* 24, lines 13ff.

13. Aristotle, *Physics* 204b13–19, 22.

14. Again, Simplicius, *Commentary on Aristotle's Physics* 24, lines 13ff.

15. The attempt to understand Anaximander's *apeiron* more clearly is something of a cottage industry. Only a brief survey can be given here, but see Charles Kahn, *Anaximander and the Origins of Greek Cosmology* (New York: Columbia University Press, 1960); H. B. Gottschalk, "Anaximander's *Apeiron*," *Phronesis* 10 (1965): 37–53; C. J. Classen, "Anaximander and Anaximenes: The Earliest Greek Theories of Change?" *Phronesis* 22 (1977): 89–102; Elizabeth Asmis, "What is Anaximander's Apeiron?" *Journal of the History of Philosophy* 19 (1981): 279–97; Aryeh Finkelberg, "Anaximander's Conception of the *Apeiron*," *Phronesis* 28 (1993): 229–56; Robert Hahn, *Anaximander and the Architects: The Contributions of Egyptian and Greek Architectural Technology to the Origins of Greek Philosophy* (New York: State University of New York Press, 2001).

16. Thus Aryeh Finkelberg, "Xenophanes' Physics, Parmenides' Doxa and Empedocles' Theory of Cosmogonical Mixture," *Hermes* 125, no. 1 (1997): 9.

17. Hippolytus *Refutations of All Heresies* 1.14.5–6. For the reliability of Hippolytus, see Jaap Mansfield, "Theophrastus and the Xenophanes Doxography," *Memnosyne* 40, no. 3–4 (1987), 290 & 293–4, Mansfield being generally favorable. Conversely, Aryeh Finkelberg, "Studies in Xenophanes," *Harvard Studies in Classical Philology* 93 (1990): 117–26 asserts that Hipploytus gives a "Parmenidized" version of Xenophanes's theology in the sentences immediately preceding the quotation here, and is more distrusting. However, he then goes on (133) to use our passage without questioning its representation of Xenophanean doctrine; in Finkelberg, "Xenophanes'," 10–13 the fidelity of our passage is defended.

18. Thus W. K. C. Guthrie, *A History of Greek Philosophy*, vol. 1 (Cambridge: Cambridge University Press, 1962), 386, and see Lesher, *Xenophanes*, 126–127.

19. Or so goes current scholarly thinking; see Finkelberg, "Anaximander's," 235–236. For the exact opposite view, namely, that the term extends the scope of its reference, see Kahn, *Anaximander*, 145.

20. Lesher, *Xenophanes*, 127, following Kathleen Freeman, *The Presocratic Philosophers* (Oxford: Blackwell, 1953), 100.

21. For the possible influence of Anaximander on Xenophanes's physics, see Finkelberg, "Studies," 112–13.

22. Thus Finkelberg, "Xenophanes'," 11: "it literally and most naturally means that in dissolving the earth the sea becomes more and more muddy until everything turns into a more of less uniform muddy solution," the question here being just how uniform we are to understand that mixture.

23. Here it seems Finkelberg is in favour of the most radical of interpretations: "the cyclicality of cosmic evolution implies that after the intermingling of the earth and the sea terminates in the complete dissolution of the world into mud, the reverse process begins, that of the breaking up of mud into two separate components" (Finkelberg, "Xenophanes'," 13).

24. Brad Inwood, ed., *The Poem of Empedocles: Fragments: A Text and Translation with an Introduction* (Toronto: University of Toronto Press, 2001).

25. T. M. Robinson, ed., *Heraclitus: Fragments: A Text and Translation with a Commentary* (Toronto: University of Toronto Press, 1987).

26. Aristotle, *Metaphysics* 985b23–986a26.

27. Simplicius, *Commentary on Aristotle's 'On the Heavens'* 294.33.

28. Aristotle, *On Generation and Corruption* 324b35–325b5.

29. David Gallop, ed., *Parmenides of Elea: Fragments: A Text and Translation with an Introduction* (Toronto: University of Toronto Press, 1984).

30. Thus Gerard Naddaf, "On the Origin of Anaximander's Cosmological Model," *Journal of the History of Ideas* 59, no. 1 (1998), 1–28, who argues that Anaximander's cosmology is "the result of political inspiration" (2). The point is also well expressed in Vernant, *Origins*, 119–27.

Chapter 2

The Immortality of the Soul and the Origin of the Cosmos in Plato's *Phaedo*

Ann Ward

In the *Phaedo* Socrates defends the desirability of his own death. To persuade the two main interlocutors of the dialogue, Simmias and Cebes, of the goodness of death, Socrates must also persuade them of the immortality of the soul. Although ultimately showing its insufficiency, Socrates first turns to Presocratic natural science, or the study of matter in motion, to demonstrate that the soul is immortal. Socrates, therefore, shows that politico-religious questions, such as whether or not the soul is immortal, will inevitably, no matter how insufficient, turn into questions of natural science, such as how and why the cosmos came into being.

Socrates' argument for the immortality of the soul developing out of Presocratic natural science begins with his first argument from contraries (69e–72e).[1] He then discusses his disillusionment with natural science to reveal the ultimate source of causality in the universe, leading to what Socrates calls his "second sailing" (95a–102a). The latter culminates in his second argument from contraries that serves as Socrates' final "proof" of the immortality of the soul (102a–107b). I argue that the core of the problem with the argument from Presocratic natural science, or the first argument from contraries, is that while the inquiry into material nature shows *that* or *how* the dead give life to the living, it does not show *when* and *why* the soul, and all other living things, initially came into being, thus allowing the soul to participate in its transmigration and, in the case of material nature, its motion. In short, although Presocratic natural science shows *how* things do what they do, it does not explain *why* things do what they do. Socrates' second argument from contraries, I argue, resolves the core problem of the argument from natural science through the Ideas as cause. In the second argument from contraries, the Ideas as cause are inseparable from their particular manifestations in the natural world. As such, they, and Socrates, cannot ensure the immortality of the individual soul, but only of the cosmos of which the soul is a part. I conclude by exploring Socrates' discussion of "misology" or the hatred of dialogue (89b–91c), and how this phenomenon can shed light on Socrates' understanding of the relation between the Ideas and our life, not simply within the natural world but also within the polis or political world.

DEATH, IDEAS, AND THE SOUL

Before turning to Socrates' analysis and then correction of the argument based on Presocratic natural science, I will briefly consider Socrates' definition of death and his explanation of the philosopher's relationship to this condition. It is this discussion of philosophy and death that initially causes the question of the immortality of the soul to be raised.

Socrates and Simmias initially agree that death is the separation of the soul from the body, in which the body comes to be "itself all by itself," and the soul likewise, once freed from the body, comes to be "herself all by herself" (64c). The philosopher has a special relationship to this condition of death because, while living, according to Socrates, he or she denigrates or rejects the body and its pleasures, such as those of "food and drink" and all "the pleasures of love-making." He or she also eschews all the material goods and luxuries that serve the body, for instance "diverse cloaks and sandals and the other, body-related beautifications" (64d–e). The philosopher turns away from satisfying the body and pursuing material goods because they interfere with the soul's attempt to "get in touch with truth" (65c–d). Moreover, the bodily senses of seeing, hearing, tasting, touching and smelling tend to conceal rather than reveal the truth to the soul in "her act of reasoning" (66a). Socrates' point is that in life, the philosopher seeks to suppress the body in order to free the soul from the body's distorting effects in its search for truth. The philosopher, in trying to separate soul from body, attempts to make life as much as possible like the condition of death.[2]

The anti-materialism of philosophy in Socrates' discussion of the desirability of death is related to what Socrates portrays as the nature of the truth that is the object of the philosophic quest. According to Socrates, the philosopher's soul, when "she comes to be herself all by herself as much as possible," strives for "what is" (65d–e). Socrates is referring to his theory of the "Ideas" or "Forms." He gives a somewhat fuller account of what he means by the Ideas later on in the *Phaedo* when, explaining his "second sailing" to Cebes, Socrates refers to those "much-babbled-about things" such as the "Beautiful Itself by Itself and a Good and a Big and all the others" (100b).[3]

Taking the Idea of the Beautiful as an example, Socrates claims that all of the particular manifestations of beauty, such as "a blossoming color or shape," are beautiful "because [they] participate in that Beautiful" (100c, d). Socrates continues and says, "nothing makes a thing beautiful but the presence of or communion with that Beautiful" and "it's by the Beautiful that all beautiful things are beautiful" (100d, e). Socrates appears to suggest that the Idea of the Beautiful is that universal characteristic in which all particular examples of beauty participate or share; it is that quality or "class characteristic" of beauty that makes the beautiful things what they are—beautiful—and thus what allows us to call them beautiful.[4] Moreover, in his discussion of the philosophic quest near the beginning of the *Phaedo,* Socrates indicates that whereas the particular manifestations of the Ideas in the material world, such as a beautiful color and shape, can be grasped by the senses of the body, the Ideas themselves, such as the Beautiful Itself, can only be grasped by reason in the soul. Socrates thus suggests that the Ideas, while in the material world inasmuch as they move through and hence make the particulars what they are, also transcend the material world in some sense (65e–66a). The anti-materialism of the philosopher, therefore, is driven by the supra-materialism, as it were, of the Ideas, the object of the philosophic quest.[5] According to Socrates, only the person "who neither puts any sight into his thinking nor drags any sense along with his reasoning" will grasp hold of the truth, and thus philosophers must "free (themselves) from the body and behold things themselves with the soul herself" (65e–66a, 66e).

Socrates concludes his argument for the desirability of his own death by reference to the Ideas. If the Ideas can only be grasped by the soul when she is completely free from the distorting effects of the body, then only after death can the philosopher attain what his or her entire life has been devoted to. According to Socrates, "the thoughtfulness we desire and whose lovers we claim to be will be ours—when we've met our end and, as the argument shows, not while we're alive" (66e). Socrates presents death as that which allows the soul of the philosopher to finally be alone with the Ideas, and thus eternally grasping and thinking the "True" (67b). Two problems arise, however, with the way in which Socrates argues for the desirability of his own death. First, when explaining to Simmias why the philosopher should not fear death, Socrates asks, "wouldn't it be great unreason if they were terrified and made a fuss when [death] happened, and weren't pleased to go There, where there was hope for those who'd arrived to get what they've been in love with throughout life—and they were in love with thoughtfulness—and to be free of the company of that with which they'd been at odds" (67e–68a). By suggesting that the philosopher, after death, is going "There," and that after he or she has "arrived" he or she will finally think the Ideas, Socrates appears to put the Ideas, and thus the soul as well, in an immaterial, supernatural place or region, separate from their particular material and natural manifestations in this world. Is this possible? For instance, we could acknowledge that the Ideas are universal, immaterial to some extent, and perhaps even eternal, but deny that they are self-subsisting or can be known wholly separate from their particular manifestations. Mustn't the Ideas always make themselves known to our reason through their particular manifestations?

The second problem with Socrates' defense of the desirability of his own death is that the image of death as that which allows the soul to finally be alone with the Ideas assumes that the soul, along with the Ideas, is immortal. Yet, Socrates has not argued for the immortality of the soul thus far, and it is clear that his interlocutors do not accept the truth of such an assumption at this point in the conversation. Cebes responds to Socrates' image of the afterlife by claiming that many people, "fear that the soul, once she's free of the body, is no longer anywhere and is destroyed and perishes on that very day when a human being dies; . . . this point—that the soul is when the human being dies and holds on to both some power and thoughtfulness—probably stands in need of more than a little persuasive talk and assurance" (70a–b). Cebes's implication is that we and the philosopher should fear death not because we love the body but because we love the soul and thinking itself, which, Cebes now suggests, perishes along with the body when we die. Cebes, therefore, does not agree that the death of the philosopher, like his or her life, is the separation of the soul from the body, but rather that it is the perishing of both soul and body. Socrates, in other words, in order to convince them of the goodness of his own death, must also convince them of the immortality of the soul.

FIRST ARGUMENT FROM CONTRARIES: NATURAL SCIENCE

In the attempt to persuade Simmias and Cebes of the immortality of the soul, Socrates begins with his first argument from contraries (69e–72e). He suggests to Cebes that they, "take a look at all things that have becoming—whether they all, as contraries, come to be from anywhere else but from their contraries, at least those that happen to have some such contrary" (70e). In considering "things that have becoming," the first argument from contraries is a turn away from the focus on the immaterial and unchanging Ideas to the material and changing particulars that come to be and pass away in the natural world. It is, in other words, a turn to Presocratic natural philosophy or "science."[6] As such, the first argument from contraries attempts to show the immortality of the soul without reference to the Ideas.[7]

Socrates' thesis is that all things come from their contraries, and all contrary things have two becomings (71a–b). For instance, the bigger comes from the littler, which is the first becoming of the contrary, and the littler comes from the bigger, which is the second becoming of the pair. The other examples Socrates gives follow this pattern, with the weaker coming from the stronger and the stronger from the weaker, the quicker from the slower and the slower from the quicker, the worse from the better and the better from the worse, the more just from the more unjust and the more unjust from the more just, and most importantly, being dead from being alive and being alive from being dead (71d–e). The latter set of contraries shows, according to Socrates, that the souls of the living come from the souls of the dead, and thus, "it's necessary for the souls of the dead to *be* somewhere, whence they come to be again" (72a). The soul, therefore, continues to exist after the death of the body and, "as if (it) were going in a circle," can enter into and give life to a new body (72b).[8]

There are two possible explanations for what Socrates means by the argument that all things come from their contraries and all contraries have two becomings. The first is that Socrates means to suggest the relativity or relatedness of all things in terms of our own perception of them. For example, if person A is six feet tall and person B is five feet tall, A beside B would appear bigger than B who would appear littler, and in this way we could say that the bigger, A, comes from the littler, B. Yet, if A stands beside person C who is seven feet tall, A would appear smaller than C who would appear bigger, and in this way we could say that the littler, A, comes from the bigger, C.[9] However, the second possible explanation for what Socrates means when he argues that contraries come from their contraries is that he is pointing to the physical causes of matter in motion.[10] This comes to light in a short exchange between Socrates and Cebes. Socrates asks:

> [B]etween a bigger and a littler thing isn't there growth and decay, and so we call the one "growing" and the other "decaying?"
> Yes, [Cebes] said.
> And separating and combining and cooling and heating and likewise everything else . . . isn't it in fact necessary that they come to be from one another and that there's a becoming from each member of the pair to the other?
> Certainly, said [Cebes] (71b–c).

Concerning what is between bigger and littler, let us take Socrates' assertion of "growth and decay" first. If, for example, a tree grows, we can say the tree changes from being littler to being bigger, and when at some point in time the tree begins to decay, we can say that it changes from being bigger to being littler, the latter ultimately culminating in death. With respect to "separating and combining," if, for example, rain water is combined with earth, we get mud and such moisture enables a tree to grow, becoming bigger, yet if water is separated from earth, we get dust or drought, which causes the tree to die, becoming "littler" or dead. Finally, with respect to "cooling and heating," if, for example, cold is added to water, we get ice, and plants die, becoming "littler," but if heat is added to ice, we get water and plants grow, becoming bigger. Socrates' suggestion, therefore, that "growth and decay," "separating and combining," and "cooling and heating" are those processes between the bigger and the littler appears to be a reference to the causes of matter in motion in the natural world. They explain what causes living things and nonliving things like earth, which can become mud, and water, which can become ice, to change or grow and decay, and therefore what causes them to be in motion, as it were.

Socrates emphasizes that in the material world "growth and decay," or coming to be and passing away, shows that what is dead gives life to what is living. Considering whether the contrary were true, Socrates asks, "if . . . when (all things that partake of living) died, the dead were to stay in that shape and not return to life, then wouldn't there be a great necessity for all things to end up dead and for nothing at all to be alive? . . . [W]hat trick would there be to prevent all things from being utterly spent in death" (72c–d)? Socrates suggests that material life on earth has not come to an end and that it will not end in the future because matter which passes away feeds or gives life to other matter which then comes to be, with this exchange or cycle repeated over and over again. Thus, for example, a tree dies and falls to the earth. Yet, its dead matter seeps into the ground and provides nourishment for a new tree to grow, the material of the dead tree acting as a fertilizer of sorts to the new tree.[11] Moreover, Socrates suggests that this principle or action of the natural, material world can be applied to the human soul. As the matter of the dead tree nourishes a new tree, so the soul of a dead person enters the body of and gives life to a new person. Thus the souls of the dead give life to the living.

There are a number of problems we should notice with Socrates' argument for the immortality of the soul based on the logic of the cycle of life in the natural world. First, it does not seem to promise the immortality of one's individual soul with one's individual consciousness. Second, it jumps from the motion of matter to the motion or transmigration of the soul, equating the behavior of the material with the immaterial.[12] Most interesting, however, is that although it purports to explain the *motion* of matter through "growth and decay," "separating and combining," and "cooling and heating," the first argument from contraries cannot explain the original *cause* of matter. For instance, although it may show *how* a dead tree can give life to a new tree, it cannot answer the question of *why* and when the first tree(s) came into being, and likewise the first coming into being of things like water and earth, hot and cold. By analogy, the argument provides no answer to the question of the origin or coming into being of the first soul(s), before its/their original transmigration began.[13] These questions ultimately point to the larger issue of whether or not the cosmos is eternal or created in time. Socrates' first argument from contraries, however, does not address this issue. Yet, it does serve to illustrate the fact that politico-religious questions, such as whether or not the soul is immortal, will inevitably turn into questions of natural science, such as why and how the cosmos came into being. Socrates suggests that political philosophy initially cannot do without natural philosophy.[14]

THE SECOND ARGUMENT FROM CONTRARIES: IDEAS AS CAUSE

The second argument from contraries addresses and attempts to resolve what I have argued is the main problem with the first argument from contraries, or the argument from Presocratic natural science. Although "growth and decay," "separating and combining," and "cooling and heating" can explain the *motion* of matter, or *how* things do what they do, they cannot explain the *cause* of matter, or *why* things do what they do, or in other words why things are what they are. In Socrates' second argument from contraries, the Ideas act as the cause of things in the cosmos, or explain why things come into being as what they are.

The Ideas are found in human speech, the turn to which Socrates' calls his "second sailing in search of the cause" (99d).[15] Socrates characterizes his "second sailing" as a turning away from the direct inquiry into "beings"—the inquiry of natural science—toward an indirect inquiry by way of studying the speeches of human beings (99d–e). According to Socrates, "it seemed to me that I should take refuge in accounts and look in them for the truth of beings" (99e). What do the speeches of human beings reveal? Socrates says to Cebes, "I'm going to try to show you the form of the cause with which I've busied myself. And I'll go back to those much-babbled-about things and take my beginning from them, putting down as hypothesis that there's some Beautiful Itself by Itself and a Good and a Big and all the others. If you give me those and grant that they *are,* I hope, from them, to show you the cause and to discover how soul is something deathless" (100b–c). Thus, human speech, according to Socrates, reveals the Ideas. Moreover, when his interlocutors grant him that the Ideas exist, Socrates proceeds to show that the Ideas are the cause of matter and its motion, or why matter is what is and does what it does, and from this that the soul is immortal. The Ideas as cause and as ensuring the immortality of the soul is what composes the second argument from contraries, Socrates' final attempt to show his interlocutors that the soul is altogether deathless (102b–107b).

In this argument from contraries, however, the Ideas do not act as the material or physical cause of things but the immaterial or "intelligible" cause of things.[16] For instance, Socrates, as we have seen, in the context of explaining what he means by the Ideas as cause argues, "if somebody should tell me why anything is beautiful by saying it has a blossoming color or shape or anything else of that sort, I bid farewell to that . . . and . . . hold . . . that nothing makes a thing beautiful but the presence of or communion with the Beautiful" (100d–e).[17] Thus, taking a beautiful flower as an example, its beauty can be explained materially or physically, such as its possession of a certain color or shape, or it can be explained, as Socrates urges, immaterially, such as its participation in or possession of the Idea of the Beautiful, that universal or "class characteristic" that all particular manifestations of beauty share.

Paul Stern helps us understand the Ideas as immaterial or intelligible causes of things when he argues that an entity "is both this particular, corporeal individual and this kind of individual."[18] Stern further maintains that for Socrates, "(t)he explanation of an entity must . . . account both for its existence as this particular individual and this *kind* of individual."[19] Thus, a particular thing as a particular thing can have a certain color and shape and other material qualities particular to it that are open to the senses, but it also shares this color and shape with other particular things that have the same or similar colors and shapes, such sharing making it and the others a kind or class of things, the latter only grasped by the mind. For example, a particular thing may have a short green stem and red petals, but having these characteristics also makes it part of a kind or "species" of things, as it were, known as flowers. Its "species" or class characteristics are its Idea, the Idea of the Flower. Taking this example one step further, the colors green and red can be part of a kind or species of things known as beautiful, thus allowing the flower to possess or participate in the Idea of the Beautiful as well. Moreover, the class characteristics of the flower separate or distinguish it from other things with different class characteristics. For instance, a particular thing with a tall brown trunk and green leaves rather than a short green stem and red petals has characteristics that make it part of a kind or "species" of things known as trees rather than flowers. Its "species" or class characteristics are its Idea, the Idea of the Tree.

Finally, not only do Ideas allow us to know why things are what they are in contrast to other things, for example that a flower is a flower and not a tree, and a tree is a tree and not a flower, but they also explain why things come into being as the things that they are. For instance, a tree dies but its matter seeps into the earth providing nourishment for a new tree to come into being. But, what determines that its matter, or any other matter, produces a new tree and not a new flower? Socrates appears to suggest that the cause is the matter's participation in or possession of the Idea of the Tree. Likewise, when a flower dies and its matter seeps into the earth, it provides nourishment for a new flower because its matter participates in or possesses the Idea of the Flower. Trees come into being as trees and flowers come into being as flowers because the matter that composes them participates in or possesses the Idea or "species" of the Tree and the Idea or "species" of the Flower.

The Ideas as the universal or class characteristics that all particular things in a class share can therefore be considered the "intelligible" cause of things because they tell us why things come into being as what they are and why they exist as different from other things in the cosmos. The Ideas allow us to know that something is, for example, a flower rather than a tree, beautiful rather than ugly, good rather than bad. Moreover, they allow us to name or call something a flower, beautiful, and good rather than a tree, ugly, and bad. The Ideas allow us to say that something is what it is. If there were no Ideas, or universal characteristics that particulars shared making them part of a class, something could be both a flower and a tree, beautiful and ugly, and good and bad at the same time. Things would keep changing, or matter would be in constant and chaotic, or unintelligible, motion. Speech, or naming, as well, would be meaningless, as one would call something a flower, beautiful, and good, but it would also be a tree, ugly, and bad. The Ideas, therefore, are not only the intelligible cause of things, allowing us to know what they are and why they are what they are, but are also the cause of rational or meaningful speech. The Ideas are not only revealed by speech but also allow speech to occur.[20]

"PROOF" OF THE IMMORTALITY OF THE SOUL

The preceding discussion of Socrates' understanding of the Ideas as the intelligible cause of things in the cosmos makes clear that the Ideas, as they were crucial in the argument for the desirability of death, re-emerge as crucial in the second and final argument from contraries. Yet, the Ideas in the latter argument differ from those in the former in significant ways. First, whereas the Ideas in the argument for the desirability of death were conceived as self-subsisting or supernatural beings capable of being known wholly separate from the material world, the Ideas in the second argument from contraries are reconceived as manifesting themselves only through their particular, material examples in the natural world. Hence, in the midst of his second argument from contraries, Socrates says, "(i)f you should ask me what comes to be in the body by which that body will be hot, I won't . . . say that it's Hotness; I'll say it's fire. Nor when you ask what comes to be in a body by which that body will be sick, will I say that it's Sickness but rather that it's fever" (105c). Socrates thereby suggests that the Ideas such as Hotness and Sickness only come to be in a body through their particular manifestations, fire in the case of Hotness and fever in the case of Sickness. Contact between the material and the Ideal worlds, as it were, causing the material to be what it is and do what it does, for example a body to be hot and to get sick, is always mediated, Socrates suggests, through the particulars that participate in or possess the Ideas. The Ideas are intelligible or allow themselves to be known only through their particulars.[21] As the Ideas are not intelligible independent of their particular manifestations, so, I argue, they do not seem to promise life to the *individual* soul beyond its material or bodily world.[22]

Socrates' second argument from contraries, or from the Ideas as cause, maintains the opposite thesis to his first argument from contraries, or the argument from Presocratic natural science. In his argument from natural science, Socrates held that all things come from their contraries, for instance the bigger comes from the littler and the littler from the bigger. Now, however, in his second argument from contraries, Socrates maintains that contraries never admit of their contraries, or in other words that contraries do not come to be from one another. Socrates explains this apparent "about-face," as it were, by arguing, "[t]hen it was said that a *contrary thing* comes to be out of a contrary thing, but now its being said that that contrary itself would never become contrary to itself" (103b). Thus, whereas in the first argument from contraries the particular manifestations of the Ideas were being discussed, in the second argument from contraries the Ideas themselves are at issue. The result of this is that when the Idea of the Small approaches the Idea of the Big, according to Socrates, "(e)ither Bigness must flee and get out of the way when its contrary, the Small, advances towards it, or else it must already have perished by the time that Smallness came near it. But what it's *not* willing to be is, by enduring and receiving Smallness, other than what it was" (102e). The Idea of the Big, therefore, never admits and thus never becomes the Idea of the Small, and the Idea of the Small never admits and thus never becomes the Idea of the Big. Contrary Ideas do not come to be from contrary Ideas; they are unchanging or simply are what they are; they are eternal.[23]

Keeping in mind that the Ideas as reconceived in the second argument from contraries only appear or show themselves in their particular manifestations,

Socrates turns to the behavior of fire, snow, the numbers three and two, and the soul. They are, respectively, particular manifestations of the Ideas of Hot, Cold, Odd, Even, and Life. He makes use of these particulars and their Ideas to illustrate his contention that contrary Ideas never admit or become contrary Ideas. I will focus on fire, snow, and soul and Hot, Cold, and Life. His "proof" proceeds in the following manner. Fire, which contains the Idea of the Hot, never admits the Idea of the Cold; although fire is not contrary to the Cold, the Hot is contrary to the Cold. If fire did admit the Cold, perhaps through the particular manifestation of water or snow, it would cease to be what it is, or would be extinguished. Likewise, snow, which contains the Idea of the Cold, never admits the Idea of the Hot; although snow is not contrary to Hot, the Cold is contrary to Hot. If snow did admit the Hot, perhaps through the particular manifestation of fire, it would cease to be what it is, or would melt. Thus, fire does not come from snow and snow does not come from fire, because the Idea of the Hot does not admit or come from the Idea of the Cold and the Idea of the Cold does not admit or come from the Idea of the Hot. The Ideas always are what they are, thus explaining the behavior of their particular manifestations.[24]

Socrates, after discussing fire and snow, moves on to the crucial question of the soul. In the following exchange with Cebes, Socrates establishes that the soul is not an Idea itself, but a particular manifestation of the Idea of Life whose contrary is the Idea of Death. Socrates asks:

> What comes to be in a body by which the body will be living?
> It's soul, [Cebes] said.
> And isn't this always the case?
> Why, of course, said [Cebes].
> Then doesn't soul always come on the scene bringing Life to bear on whatever she herself occupies?
> Of course that's how she comes on the scene, [Cebes] said.
> Is there some contrary to Life . . . ?
> There is, [Cebes] said.
> What?
> Death (105c–d).

Socrates suggests that the soul is a particular manifestation of the Idea of Life whose contrary is Death. Thus, following the previous pattern of argumentation, Socrates says that the soul, which contains the Idea of Life, will never admit the Idea of Death, although not contrary to it. As the following exchange suggests, Socrates thereby appears to prove the immortality, or "undying" character, of the soul.

> [W]hat do we call whatever doesn't admit Death?
> Un-dying, [Cebes] said.
> And soul doesn't admit Death?
> No.
> Therefore soul is something un-dying?
> Something un-dying.
> Alright . . . shall we claim that this has been demonstrated?
> Very sufficiently demonstrated indeed, Socrates (105e).

Socrates' "demonstration" of the immortality or "imperishability" of the soul based on the argument that, because it contains the Idea of Life, it never admits the Idea of Death is problematic when we consider more closely the behavior of snow and fire, two analogies to the soul Socrates uses to illustrate his argument.[25] Socrates argues, "(i)f it were necessary that the un-hot be imperishable as well, whenever anybody brought Hot upon snow, wouldn't snow slip away safe and unmelted? For surely it wouldn't have perished, nor again would it have endured or admitted Hotness . . . And in the same way, I imagine, if the uncold were imperishable as well, whenever anything cold came at fire, it would never be extinguished or perish, but would take off and go safe and sound"(106a–b). Socrates' argument in this passage is quite strange. Anyone who has lived through a raging winter knows and is glad that when the warm air of spring finally arrives, the snow does not "slip away safe and unmelted," but rather melts and disappears. Likewise, when I dowse the fire in my fireplace with water, it doesn't "take off and go safe and sound," but is, hopefully, extinguished. Perhaps what Socrates means to suggest in this passage is that while a particular patch of snow may melt and therefore perish when spring comes, more snow will come again next winter. Snow itself, as a phenomenon, doesn't cease to exist in the world because the Idea of the Cold, of which snow is a particular manifestation, is "un-dying" or eternal. Likewise, although a particular fire can be extinguished by dowsing it with water, we can always light another one the next evening. Fire itself, as a phenomenon, will not disappear altogether from the world because the Idea of the Hot, of which fire is a particular manifestation, is eternal.

What does the behavior of snow and fire imply about the soul? Perhaps a particular soul can die or "scatter," but soul itself as a phenomenon will not disappear from the world because the Idea of Life, of which soul is a particular manifestation, is eternal. Thus, while the individual soul of a particular person may die, other souls will come to be and continue to live, now and always. The second argument from contraries does not seem, therefore, to demonstrate the immortality of the individual soul, but rather the immortality of the cosmos in which the individual soul comes to be, lives, and then passes away.[26] Socrates, looking to the Ideas as "cause," holds out the hope of a stable universe in which the character of human life as we know it and the natural environment of which it partakes will always exist.[27]

IMPLICATION FOR POLITICS

What are the implications of Socrates' first argument from contraries and its subsequent failure for politics? As we have seen, although it can explain the *motion* of matter, or *how* things do what they do, the first argument from contraries cannot explain the *cause* of matter, or *why* things do what they do, or in other words, why things are what they are. By analogy, the argument provides no account of the origin or coming into being of the first soul(s) before its/their original transmigration began. These questions ultimately point to the larger issue of whether or not the cosmos is eternal or created in time. Thus, the first argument from contraries serves to illustrate the fact that human questions, such as whether or not the soul is immortal, will inevitably turn into questions of natural science, such as why and how the cosmos came into being. Socrates, therefore, suggests that political philosophy initially cannot do without natural philosophy.[28]

Socrates' first argument from contraries, however, although pointing toward it, fails to address the larger issue of the origin in time versus the eternity of the universe. Such insufficiency leads Socrates to his "second sailing in search of cause" and his subsequent discovery of the Ideas or Forms. Yet, although the Ideas apparently show the eternity of the universe, they cannot show the immortality of the individual soul, that for which politics initially began its search. Thus, in what way do the Ideas inform Socrates' thinking about politics? Perhaps some brief reflections on a short digression in the middle of the dialogue in which Socrates warns against misology, or the hatred of speeches, with which I will conclude, can shed light on this question.

Phaedo, the narrator of the dialogue, tells us that Socrates warned his friends against misology, or the distrust of speeches, claiming, "it's not possible . . . for anybody to experience a greater evil than hating arguments" (89d).[29] The cause of misology is illustrated by Socrates through an analogy, misanthropy or the hatred of human beings. Misanthropy, according to Socrates, arises when one of our heroes, someone whom we admire and think is truthful, is debunked when we learn, usually against our will, that they are liars or not who we thought they were. When this happens repeatedly, we come to hate all human beings, believing every person we meet is "wicked and untrustworthy" (89d–e). Likening speeches to human beings, Socrates argues that misology arises when we are persuaded of the truth of an argument, for example for the immortality of the soul, which is later debunked. According to Socrates, when this happens repeatedly, instead of blaming our own ineptitude or ignorance, we blame speeches or arguments for not being able to reveal the truth. This is the greatest of evils because, "finish(ing) out the rest of his life hating and reviling arguments," the misologist is "robbed of the truth and knowledge of the things that *are*" (90b, d).

Socrates claims that misology is most likely to affect those "who've spent their days in debate-arguments." He is referring to the sophists who come to believe "that they alone have detected that there's nothing sound or stable—not in the realm of either practical matters *or* arguments—but all things that *are* simply toss to and fro . . . and don't stay put anywhere for any length of time" (90c). The perspective of the sophist, which also appears similar to the perspective of Presocratic natural science, is that all reality is simply chaotic motion. Speech, which is supposed to reflect or signify this reality, is also in constant motion as words keep changing their meaning or have no stable meaning at all. Thus, what is true for you is not true for me because there is no unchanging or universal truth that exists beyond the perspective of the speaker.

I would suggest that the only thing preventing the perspective of sophists from being true are the Ideas. The Ideas ensure, for instance, that fire, which contains the Idea of the Hot, is not at the same time snow, which contains the Idea of the Cold, and thus also ensure that my words, when I say that this thing is fire which is hot, cannot also mean that this thing is snow which is cold (103c–d, 104e–105a).[30] Similarly, they ensure that my words, when I say, for instance, that this thing is a flower, beautiful, and good, cannot also mean that this thing is a tree, ugly, and bad. The Ideas, in other words, make rational or meaningful speech possible. Thus, if we accept Aristotle's claim that "man is by nature a political animal" because "[he] alone among the animals has speech" which "serves to reveal . . . good and bad, just and unjust" (1253a2–4, 9, 14–17), then it seems that from a Socratic perspective, it is the Ideas that make such a discursive understanding and practice of politics articulated by Aristotle possible.[31] The Ideas, therefore, ground not only the cosmos but also the polis, and thus, if not able to show the fate of the individual soul in the next life, can show how it flourishes in this life.

ENDNOTES

1. Plato. *Phaedo*. trans. Eva Brann, Peter Kalkavage, Eric Salem, trans. (Newburyport, MA: Focus Classical Library, 1998). All subsequent citations will be taken from this edition.

2. See Lloyd Gerson, *Knowing Persons: A Study in Plato* (Oxford: Oxford University Press, 2003), 55, 57–58; Christopher Bobonich, *Plato's Utopia Recast: His Later Ethics and Politics* (Oxford: Oxford University Press, 2002), 14–15; Seth Bena-

rdete, *The Argument of the Action: Essays on Greek Poetry and Philosophy* (Chicago: University of Chicago Press, 2000), 288; and David Bostock, *Plato's Phaedo* (Oxford: Oxford University Press, 1986), 25, 27, 29; but see Coleen Zoller, "Seducing Socrates and Resolving Plato's Separate Soul Paradox," in *Socrates: Reason or Unreason as the Foundation of European Identity*, ed. Ann Ward (Newcastle-upon-Tyne: Cambridge Scholars Publishing, 2007), 30–44; Lisa Pace Vetter, *"Women's Work" as Political Art: Weaving and Dialectical Politics in Homer, Aristophanes, and Plato* (Lanham, MD: Lexington Books, 2005), 135–137, 143, 147; and Daniel C. Russell, *Plato on Pleasure and the Good Life* (Oxford: Oxford University Press, 2005), 78, 90–92, 100–102.

3. See Ronna Burger, *The Phaedo: A Platonic Labyrinth* (New Haven, CT: Yale University Press, 1984), 7.

4. Also see Plato, *Republic*, trans. C. D. C. Reeve tran (Indianapolis: Hackett Publishing Company, 2004), 476a–b, 479a; Paul Stern, "Antifoundationalism in Plato's *Phaedo*," *The Review of Politics* 78, no. 3 (Spring 1989): 203; Mary P. Nichols, *Socrates and the Political Community: An Ancient Debate* (Albany: State University of New York Press, 1988), 112; and Bostock, *Plato's Phaedo*, 147, 207; but see F. C. White, "The *Phaedo* and *Republic V* on Essences," *The Journal of Hellenic Studies* 98 (1978): 142, 143, 148.

5. Gerson, *Knowing Persons*, 59–61; Bobonich, *Plato's Utopia*, 25–26, 28.

6. See Bostock, *Plato's Phaedo*, 141; Michael Davis, "Socrates' Pre-Socratism: Some Remarks on the Structure of Plato's *Phaedo*," *Review of Metaphysics* 33, no. 3 (March 1980): 562, 566; and C. C. W. Taylor, "Forms as Causes in the *Phaedo*," 78, no. 309 (January 1969): 45.

7. But see Gerson, *Knowing Persons,* 63; and Stern (1993), 50–51.

8. But see Gerson, *Knowing Persons*, 64; Peter J. Ahrensdorf, *The Death of Socrates and the Life of Philosophy* (Albany: State University of New York Press, 1995), 68; and Bostock, *Plato's Phaedo*, 42–43, 52–53, 58.

9. But see Bostock, *Plato's Phaedo*, 48.

10. But see Paul Stern, *Socratic Rationalism and Political Philosophy: An Interpretation of Plato's Phaedo* (Albany: State University of New York Press, 1993), 51, 56, 60.

11. See Kenneth Dorter, *Plato's Phaedo: An Interpretation* (Toronto: University of Toronto Press, 1982), 36; but see Bostock, *Plato's Phaedo*, 54–56.

12. Davis, "Socrates' Pre-Socratism," 568.

13. See Gerson, *Knowing Persons*, 51; Stern, *Socratic Rationalism*, 55, 159; and Davis, "Socrates' Pre-Socratism," 568.

14. See Stern, "Antifoundationalism," 201.

15. For Socrates' disappointment with Anaxagoras and his resultant insistence on teleology to explain the motion of matter or *why* rather than how matter does what it does, see Plato, *Phaedo*, 97c–99c.

16. Gerson, *Knowing Persons*, 56, 93; Stern, "Antifoundationalism," 196, 203–204, 205; Taylor, "Forms," 46, 50.

17. See Davis, "Socrates' Pre-Socratism," 563–564.

18. Stern, *Socratic Rationalism*, 197.

19. Stern, *Socratic Rationalism*, 198.

20. See Bostock, *Plato's Phaedo*, 148.

21. See Bostock, *Plato's Phaedo*, 185–187, 199.

22. See Ahrensdorf, *Death of Socrates*, 134; Stern, "Antifoundationalism," 194; and Dorter, *Plato's Phaedo*, 159.

23. But see Ahrensdorf, *Death of Socrates*, 175, 176, 187; and Stern, *Socratic Rationalism*, 148, 153, 157–158 and "Antifoundationalism," 202.

24. But see Taylor, "Forms," 50, 51, 52.

25. See Ahrensdorf, *Death of Socrates*, 182–183; Stern, *Socratic Rationalism*, 148, 154–155, 160–161 and "Antifoundationalism," 206; Bostock, *Plato's Phaedo*, 189–192; and Dorter, *Plato's Phaedo*, 151–153, 157–161, 181.

26. But see Gerson, *Knowing Persons*, 95–96; and White, "The *Phaedo*," 310.

27. But see Stern, *Socratic Rationalism*, 161 and "Antifoundationalism," 204, 205, 206.

28. See Stern, "Antifoundationalism," 201.

29. See Benardete, *The Argument*, 279; and Burger, *The Phaedo*, 9–10.

30. See Benardete, *The Argument*, 295–296; and Bostock, *Plato's Phaedo*, 161, 208–209.

31. Aristotle, *The Politics*, Carnes Lord trans. (Chicago: University of Chicago Press, 1984).

Chapter 3

Plato's Science of Living Well

Coleen Zoller

While science, mathematics, and philosophy are for the most part distinct disciplines today, in the fourth century BCE they were regarded collectively as philosophy. Plato considers a mathematical-scientific understanding of the cosmos integral to the philosopher's love of wisdom. He does not merely think that "doing science" will improve our understanding of the cosmos; rather, he believes that looking at life with an eye toward mathematics and science improves our understanding of how human beings should live if they want to flourish at being human.

It may strike us as a paradox for Plato to hold that understanding mathematics will make one a better decision-maker in all facets of life. As Myles Burnyeat asks, "How, we may ask, will knowing how to construct an icosahedron help them when it comes to fixing the price of fish in the ideal market or understanding the Platonic Theory of Forms?"[1] Yet, Plato does take quite seriously the value of mathematical-scientific understanding. For instance, in the *Philebus* he explains that anyone who has "never worked out the amount and number of anything at all" will "count for nothing and amount to nothing" (17e). To this end Plato also writes in the *Protagoras* that "the art of measurement" will "save our life" (356d–e). The central aim of this paper is to explore the role of mathematical-scientific understanding in living well as an individual and as a citizen.

THE HEALTHY CITY

Even though Plato focuses his attention primarily on justice, it is fair to construe his overall project as the quest for peace, which he contends will be manifest wherever human beings are provided for in a just and moderate fashion (*R.* 372a–d). I believe Plato is right to claim that when human beings are provided for in a just and moderate fashion "they'll live in peace and good health, and when they die at a ripe old age, they'll bequeath a similar life to their children" (*R.* 372ad). Plato contends that justice can only materialize when individuals and *poleis* govern themselves in a healthy manner. Thus, there is no hope for peace (personal or political) without justice, which entails the health that follows from moderate self-governance. While Plato does not explicitly connect the quest for peace with the art of measurement, we can now see that peace can only occur in an environment where citizens use the art of measurement to calculate the impact of their lifestyle on both the health of oneself as well as of the rest of the world's citizens.

Plato often depicts the trouble human beings have engaging in this sort of reflection. As the dialogues tell it, most of us focus single-mindedly on our bodies' needs and desires. Plato construes the single-minded focus on the body, which he calls body-love (*Phd.* 68b–c), as a psychic disease. Individuals plagued by body-love value only pleasure and wealth and are in need of what Plato calls "psychic medicine." One of the symptoms of this disease is the aforementioned failure to reflect on the consequences of one's lifestyle not only on oneself but also the global network of political associations. Plato construes philosophy as psychic medicine because loving wisdom includes precisely this sort of reflection.

Often Plato is construed as believing that asceticism is the appropriate cure for psychic disease.[2] The ascetic interpretation of Plato has long been convention because there are aspects of the dialogues that sound as though they endorse asceticism. One need only think of the *Phaedo* to call most of them to mind. Although ascetic dualist interpreters imagine that Plato recommends asceticism as the cure for psychic disease, the dialogues make clear that the antidote to body-love is loving wisdom rather than loving only pleasure and wealth. Furthermore, one must learn how to be properly disposed toward the body before one can love wisdom and flourish as a human being. The central question is, then, whether loving wisdom requires an ascetic lifestyle or a moderate one. My contention is that, according to Plato, one flourishes as an individual human being when one learns how to care for the soul *without* disregarding the body's needs and desires, rather than by learning how to deny the body's needs and desires. (Elsewhere I have argued against the ascetic interpretation of the *Phaedo* as well as of the *Symposium.*[3]) Likewise, an individual *polis* and the global network of political associations flourish when neither the extreme of hedonistic indulgence ("body-love") nor of ascetic body-denial prevails.

For example, in Book II of the *Republic* Socrates indicates that the citizens of the healthy city enjoy erotic experiences with each other. Plato has

Socrates say, "They'll enjoy sex with one another but bear no more children than their resources allow, lest they fall into either poverty or war" (372b–c). These citizens enjoy the pleasures of *erôs* but regard justice (rather than physical pleasure) as their chief concern. So, as a result, they need to know how to calculate properly how many children their resources can support. Peace continues to elude us precisely because this sort of reflection about the consequences of one's lifestyle on ourselves and the rest of the world's citizens is so rare.

Assuming that human beings are capable of such reflection, Plato has Socrates set forth in *Republic* II a principle to promote justice. He says that human beings should include in their lives only what is necessary; he calls for a moderate rather than luxurious lifestyle. Yet, Plato makes clear that the appetite for a just and moderate lifestyle does not come easily to human beings. He has Glaucon interrupt Socrates to say that the moderate lifestyle sounds good enough only for pigs (*R.* 372d). Glaucon deems luxury to be the only lifestyle fit for human beings. To this Socrates replies:

> All right, I understand. It's not merely the origin of a city that we're considering, its seems, but the origin of a luxurious (*truphôsan*) city. And that may not be a bad idea, for by examining it, we might very well see how justice and injustice grow up in cities. Yet the true city, in my opinion, is the one we've described, the healthy (*hugiês*) one, as it were. But let's study a city with a fever (*phlegmainousan*), if that's what you want (*R.* 372e).

Here we see that even in antiquity there was a declaration that human beings ought to live luxuriously rather than moderately. Annas contends that "Socrates's acquiescence to Glaucon's demand is in effect a recognition that people will always go on to demand unnecessary gratifications."[4] I disagree with Annas about the tone of Socrates's response to Glaucon. Socrates's own preference for the lifestyle in the healthy city suggests that it is possible for human beings to transcend the body-love that attracts us to luxurious living. However, I will concede that putting the claim that the moderate city is the true city in the mouth of Socrates rather than one of the middling interlocutors demonstrates the extraordinary nature of people who transcend body-love and the concomitant enjoyment of luxurious living.

Given, then, how difficult it is to prefer moderation to excess, it should not surprise us that Plato devises a strategy to put his body-loving interlocutors and readers in a position to grant the importance of living the just and moderate life. He frequently uses what cognitive scientists now call analogical reasoning to make his strange views more conceivable. An analogy is the systematic relationship between two analogs, a "target" analog and a "source" analog. The target analog represents the new situation that one is trying to figure out, and the source analog represents the old experience that is being adapted and applied to the target analog.[5] Analogical reasoning relies on the ease with which one can adapt and apply one's knowledge of the source analog to the target analog. Plato relies on the body as a source analog in hopes of expounding his claims about the soul, which is his target analog.

Plato chooses to characterize the difference between the cities Glaucon and Socrates envision in terms of health and disease because he knows that human beings typically value physical health as inherently good. No one wonders why someone prefers physical health to sickness. Knowing how instinctive the concepts of physical health and disease are to his audience, Plato crafts the analogy between justice and health for the sake of making more plausible his message that when one lives immoderately (as a result of not properly caring for the soul), both the individual and the *polis* suffer from the ensuing injustice.

The healthy city becomes feverish once commodities that are unnecessary for a healthy life like perfumes, pastries, prostitutes, and other such luxuries are incorporated (*R.* 373a). For example, he explains to Glaucon that if the citizens are going to eat meat, then they will need more doctors for the individual citizens as well as an army for the city to seize more land for pasture from their neighbors. And if their neighbors "too have surrendered themselves to the endless acquisition of money and have overstepped *the limit of their necessities,*" then war will be inevitable. Socrates then says, "We won't say yet whether the effects of war are good or bad but only that we've now found the origins of war. It comes from those same desires that are most of all responsible for the bad things that happen to cities and the individuals in them" (*R.* 373c–e). It is difficult for Glaucon to understand why the inclusion of such things as couches and meat cause such problems for the city (*R.* 372c–e). Although the *Republic* does not offer an item-by-item account of why these commodities bring disease to the city, *Gorgias* 464b–465d clarifies why these inclusions are problematic for an individual. Let's now turn to that passage.

THE COMPLETE CARE OF SELF

Socrates also utilizes analogical reasoning in the *Gorgias* in order to assert the existence of expertise in the field of psychic care. Socrates notes that just as there are crafts that combine to make the body healthy, so too are there two crafts that work together to make the soul healthy. While he rests with calling the expertise about physical health the care of the body, Socrates gives the field of psychic care a technical term: politics (464b). Here Plato has Socrates indicate the broad nature of politics. We tend to use the word "politics" to refer narrowly to public governance, but in the *Gorgias* Socrates steers us away from this narrow understanding of politics toward a sense of politics that includes self-governance.

Gorgias 464b–465d surveys all that is necessary for healthy self-governance as well as how authentic self-governance (the care of the body and the soul) can be imitated. Socrates says that the care of the body is, on one hand, gymnastics and, on the other, medicine (*Grg.* 464b). He goes on to say that these are imitated by cosmetics and pastry-baking, respectively (*Grg.* 465c). The cosmetic knack is the shortcut to appearing healthy just as confections merely imitate medicine's power to make someone feel better. Resting on the couches of the *Republic*'s luxurious city is the antipode of the exercise that Socrates refers to as gymnastics in the *Gorgias,* and eating the pastries Glaucon demands is exactly what Socrates cautions us against in this passage in the *Gorgias.*

Analogously Socrates claims that politics consists, on one hand, of legislation and of justice on the other. Just as one engages in physical training (what Plato calls gymnastics) in order to prevent physical disease, one must engage in psychic training (what Plato calls legislation) in order to prevent psychic disease. And just as one submits to medical treatment in order to correct physical disease, one must submit to psychic medicine (what Plato here calls justice) in order to correct psychic disease. Furthermore, just as the healthy appearance caused by practicing gymnastics can be mimicked by cosmetics, so too can a healthy psychic appearance be given off by one who has mastered sophistry. This is exactly what Adeimantus had in mind when he worried that one need not actually possess a well-ordered soul (that is, a just soul) so long as one learns how to use persuasion to convince others that one is just (*Rep.* 365b–d). Finally, just as pastry-baking can mimic the way that medicine makes someone sick feel better, so too does oratory mimic the way that justice actually corrects psychic illness.

At best a body-lover concentrates on fitness and nutrition (and medicine too whenever necessary), but at worst a body-lover uses cosmetics to appear healthy and eats pastries to feel better physically. Regardless of which type of body-lover one has been, the body-lover who comes to see the necessity of devoting oneself to the care of the soul makes a transition toward politics as Socrates broadly construes that endeavor (namely, as psychic care). At that point one must either exercise one's own psychic legislation or submit oneself to a psychic trainer. However, if one requires the guidance of a psychic trainer, ideally the psychic trainer would teach one how to legislate one's own psychic aspects. The ideal psychic trainer is someone who is always informed by what is best rather than by what is pleasurable. Socrates defines a true politician exactly in these terms (*Grg.* 521d) as he reminds us how rare such psychic trainers are!

And whenever the soul has fallen into need of psychic medicine (due to either bad legislation or bad circumstances) the soul-lover must either carry out justice on one's own or submit to psychic correction, such as punishment through the legal system. The body-lover becomes a soul-lover by putting the psychic crafts into practice. When the psychic crafts are put into practice the soul will be in good shape, just as when one implements the physical crafts, the body will be in good shape. Practiced together this is Plato's science of living well as an individual.

In the *Gorgias* Plato has Socrates demonstrate that this healthy life entails learning how to rule oneself and be moderate in satisfying one's appetites. The interlocutors, especially Callicles, find the notion of "ruling oneself" extremely foreign. In response to Socrates's declaration that self-rule matters as much and more than ruling over others, Callicles literally asks Socrates, "What do you mean by one who rules himself?" (*Grg.* 491d). By way of setting out to show the benefits of moderation and self-rule in contrast with hedonism Socrates makes two central points. First, at 493d–494a he indicates that, while one might imagine that the hedonistic life is pleasurable, it is empirically a life of pain and frustration. Second, at 505a–b he notes that, while one might consider the hedonistic lifestyle one entirely free from constraint, it is marked by the constraints imposed by medical doctors and psychic trainers like Socrates. Let's now turn to these passages.

THE JAR METAPHOR

In the *Gorgias* Callicles balks at the idea that not all desires for pleasures ought to be satisfied (491e–492a). He claims to believe that pleasure and goodness are identical (492c), and upon hearing Callicles express this view, Socrates remembers having heard someone compare the appetitive aspect of the soul to a jar (493a). In hopes of convincing Callicles "to choose the orderly life, the life that is adequate to and satisfied with its circumstances at any given time" (493c) Socrates compares the hedonistic lifestyle to a leaky jar (*Grg.* 493b) and contrasts the leaky jar with a jar kept full by a tight lid. He says:

> Consider whether what you're saying about each life, the life of the self-controlled man and that of the undisciplined one, is like this: Suppose there are two men, each of whom has many jars. The jars belonging to one of them are

> sound and full, one with wine, another with honey, a third with milk and many others with lots of other things. And suppose that the sources of each of these things are scarce and difficult to come by, procurable only with much toil and trouble. Now the one man, having filled up his jars, doesn't pour anything more into them and gives them no further thought. He can relax over them. As for the other one, he too has resources that can be procured, though with difficulty, but his containers are leaky and rotten. He's forced to keep filling them, day and night, or else he suffers extreme pain (*Gorgias* 493d–494a).

Here Socrates rejects the hedonist lifestyle because it is difficult and painful to maintain. In contrast, he commends the lifestyle symbolized by a full jar with a tight lid. The question is: does this full jar with a tight lid symbolize the life of moderation or asceticism?

The contents Plato chooses for the jars possessed by the self-controlled person are the first indication that this metaphor is not intended to recommend the austere life of asceticism. Plato chooses wine, milk, and honey as the contents of the jars, and these choices are absolutely not arbitrary. Wine, milk, and honey are traditionally associated with Dionysus, and consequently they would be ineffective symbols of asceticism. Furthermore, if Socrates were trying to encourage the soul-lover to deny the body altogether, as ascetics do, then it strikes me as rather ineffective to represent the ascetic lifestyle as a *full* jar with a tight lid. An *empty* jar would be a much more effective representation of the ascetic denial of appetites. The point of the jar metaphor passage is, rather, to instruct us to be moderate about the satisfaction of our desires so as to be self-sufficient.

DIFFERENT PRESCRIPTIONS

Having noted above that Plato's corpus does include occasional ascetic-sounding remarks, we are left to wonder why Plato includes in his dialogues two sets of messages about whether or not the wisdom-lover's life can properly include the enjoyment of moderate pleasure. The tension between these two positions is not due to Plato being conflicted or indecisive about how he thinks an embodied inquirer should treat the body. Instead, these seemingly mixed messages reflect Plato's attempt to have Socrates write two different prescriptions for psychic health.

At 505a, Socrates points out that a doctor does not give the same prescription for health to all patients. He says, "doctors generally allow a person to fill up his appetites, to eat when he's hungry, for example, or drink when he's thirsty *as much as he wants to when he's in good health,* but when he's sick they practically never allow him to fill himself with what he has an appetite for" (*Grg.* 505a; emphasis added). The doctor need not ever restrict the sort of patient who has a scientific understanding of how much and what to eat and drink, while the doctor must rein in a patient who has not mastered the art of measurement. Once Callicles agrees to this, Socrates adds, "And isn't it just the same way with the soul . . . ? As long as it's corrupt, in that it's foolish, undisciplined, unjust and impious, it should be kept away from its appetites and not be permitted to do anything other than what will make it better" (*Grg.* 505b). Callicles agrees and soon thereafter realizes that his agreement refutes his initial claim that it is better for the soul to be undisciplined rather than disciplined. Although Callicles does not remain a cooperative interlocutor, it is clear that the body-soul analogy has done its work. It made Socrates's claim (that it is better for the soul to be disciplined in satisfying its appetites) more intuitive for Callicles. In fact, the use of analogical reasoning made Socrates's call for moderation so intuitive that it led Callicles to refute himself (505b–c).

The fact that a different prescription is given to the healthy patient and a much more restrictive prescription to the diseased patient is of great analogical significance for one trying to interpret Plato's messages about how a soul-lover ought to treat the body. If different prescriptions are written for the healthy and the sick, then one should expect to see not one but two sets of messages in Plato's corpus about how to live with respect to the satisfactions of one's desires. And indeed the reader will find two sets of messages about how to treat the body.

For his body-loving patients Socrates writes a prescription that calls for caring for the soul rather than centrally valuing the body. It is to this patient that the ascetic-sounding remarks are directed. Asceticism, denial of the body, is the antipode of hedonistic body-love, and as such it is the required first step in curing the psychic illness that inevitably results from inattention to the care of the soul. Just as a sick patient is given a restrictive prescription, so too is the patient riddled with psychic illness in need of an austere prescription. Asceticism is the strictest prescription, namely, what Socrates describes as not being "permitted to do anything other than what will make it better."

However, someone who has already made the transition away from body-love toward soul-love is not suffering from psychic illness. As a result, Socrates writes a different prescription for this patient. This patient gets a more permissive prescription. Just as a doctor allows a healthy man "to fill up his appetites . . . *as much as he wants to,*" so too is the psychically healthy patient allowed to satisfy his or her appetites as much as he or she wants to. There is no danger in allowing psychically healthy people to satisfy their appetites because healthy people have healthy desires. Healthy patients know how to be moderate because they see beyond the mere appearances of what they want, knowing instead which pleasures and how much pleasure to include in a healthy life. In the terms of the *Republic,* they have healthy desires as a result of having worked to put the soul's appetitive and spirited aspects under the guidance of the soul's rational aspect. Their health results from continued success with satisfying the right sorts of desires, namely, the ones approved of by the soul's rational aspect.

In the three main passages discussed here we have seen why Plato believes that a mathematical-scientific understanding of the cosmos brings us face to face with the truth about the consequences of our chosen lifestyle. The failure to master the art of measurement leads to psychic and physical disease as well as political conflict. This art of measurement is the underpinning of both just political interactions as well as the care of the self. When one masters the art of measurement one is in a position to make the best decisions about how to live. It will serve us well today to take a fresh look at Plato's suggestion that scientific understanding will enable one to live well both as an individual and as a citizen of the world.

ENDNOTES

1. M.F. Burnyeat, "Plato on Why Mathematics Is Good for the Soul," in *Mathematics and Necessity*, ed. Timothy Smiley (Oxford: Oxford University Press, 2000), 2.

2. E.R. Dodds, *Plato, Gorgias* (Oxford: Clarendon Press, 1959), 299; Pierre Hadot, *What Is Ancient Philosophy?* trans. Michael Chase (Cambridge, MA: The Belknap Press of Harvard University Press, 2002), 67–8, 159; Alexander Nehamas, *The Art of Living: Socratic Reflections from Plato to Foucault.* (Berkeley: University of California Press 1998), 161; W.K.C. Guthrie, *A History of Greek Philosophy.* vol. 4, *Plato: the Man and His Dialogues Earlier Period* (Cambridge MA: Cambridge University Press, 1975), 331; Paul Friedländer, *Plato,* vol. 2, *The Dialogues: First Period,* trans. Hans Meyerhoff (New York: Bollingen Foundation, 1964), 54; Paul Friedländer, *Plato*, vol. 3, *The Dialogues: Second and Third Periods,* trans. Hans Meyerhoff (New York: Bollingen Foundation, 1969), 42; L.P. Gerson, "Platonic Dualism," *The Monist* 69, 3 (1986): 352–69; K.J. Dover, "Classical Greek Attitudes to Sexual Behaviour," in *Women in the Ancient World: The Arethusa Papers,* edd. John Peradotto and J.P. Sullivan (Albany: State University of New York Press, 1984), 154; Kenneth Dorter, "The Soul's Mediation between Corporeality and the Good," in *The Classics of Western Philosophy*, eds. Jorge J.E. Gracia, Gregory M. Reichberg, and Bernard N. Schumacher (New York: Blackwell, 2003); Eduard Zeller, *Outlines of the History of Greek Philosophy* (Cleveland: Meridian Books, 1931), 141; G.M.A. Grube, *Plato's Thought* (Boston; Beacon Press, 1935), 63; Benjamin Jowett, *The Dialogues of Plato* (New York: Random House, 1892), 161; R.D. Archer-Hind, *The Phaedo* (London: MacMillan, 1894), xiii; A.E. Taylor, *Plato: The Man and His Work* (New York: Meridian, 1956), 182.

3. Coleen Zoller, "Seducing Socrates and Resolving Plato's Separate Soul Paradox," in *Socrates: Reason or Unreason as the Foundation of European Identity*, ed. Ann Ward (Newcastle: Cambridge Scholars Press, 2007).

4. Julia Annas, *An Introduction to Plato's Republic* (Oxford: Oxford University Press, 1981), 77.

5. Paul Thagard, *Mind: Introduction to Cognitive Science* (Cambridge, MA: MIT Press, 1996), 78.

Chapter 4

Understanding Aristotle's *Politics* through Form and Matter

Mostafa Younesie

Consideration of Aristotle's *Politics* as a treatise brings to light the basic fact of its composition and structure. As a treatise it has specific qualities that we have to bear in mind. If we take his treatise as the extant writings of Aristotle that intended for the use of students they are: "in the way of memoranda (not notes) which may possibly have been written afterwards in order to preserve a record of the result attained in the lectures and discussions."[1] As a treatise its structural and compositional qualities become important and have results for reading it.

Politics as a treatise is composed of different memoranda that do not necessarily have a reasonable or chronological homogeneity. Focusing only on the reasonable side of the issue, we want to answer the Barker question: Can we accept it in this form and if no what we can do for understanding it? In the other words, "How could a modern thinker arrange these (six) sections, if there were no existing order, and he were given a free hand to create an order?"[2] It seems that if we want to read Aristotle as he would have read himself and also understand his explanation of politics as a science, we cannot take all things for granted and as obvious. In this context there are various accounts proposed about the appropriate order of the structure and composition of these essays if they are to form a homogenous, reasonable unit.

Let's suppose that we have to form one written "body" out of the different but related memoranda of the *Politics*. It seems that each essay has both matter of which they consist and also an essential essence and nature. Therefore, with simultaneous attention to these two aspects we can form a written unit or body and, accordingly, the constructed body will have form and matter too. Thus the memoranda consist of some matter but this is not enough and there should be something more that is their essence. They have a kind of confluence, and we cannot study each in abstraction from the other one. If true, this heuristic supposition will direct us toward a kind of "physics of politics."[3] For, if we are to order the various and different essays of Aristotle's *Politics* into a homogeneous unit, we have to consider them in light of matter and form, concepts that are borrowed from his *Physics,* book 2. Thus, through the "physics of politics" we want to heuristically order the essays or chapters of *Politics* by defining which essays in connection with each other (not in itself) matter and which one form. Moreover, as "form is the moving or efficient cause of the coming into being and behaving of the thing; and the form is also the 'end' wherein its nature is fully realized," it is natural that for starting the reconstruction process of *Politics,* we need some material that is presented by Aristotle himself in the form of specific memoranda within and not without.[4] But the things in question cannot be there unless the related immanent materials receive the form or characteristics of the type, and its conformity to which brings it within the definition of the thing we say it is. The material is thereby arrived and realized, its being what it had to be, for it is not possible to give any form to the related materials that we want. But we have not realized the thing or entity unless there is something to start the process. In the other words and more generally, there must be the prime agent that produces the effect and starts the material on its way to the product, changing it from what it was to what it is to be. And lastly, there is the end or purpose for the sake of which the process initiated (and it is Aristotle's reading and conception of the end).[5] According to Aristotle:

> τουτων δε τα μεν ωσ το υποκειμενον (οιον τα μερη) , τα δε ωσ το τι ην ειν αι – το τε ολον και η συνθεσισ και το ειδοσ. το δε σπερμα και ο ιατροσ και ο βουλευσασ και ολωσ το ποιον, παντα οθεν η αρξη τησ μεταβολησ η στασεωσ η κινησεωσ. τα δ ωσ το τελοσ και ταγαθον των αλλον. το γαρ ου εϖεκα βελτιστον και τελοσ των αλλων εθελει ειvει (διαφερετων δε μηδεν ειπειν αυτο αγαθον η φαινομενον αγαθον). (but of these things some are causes in the sense of the substratum (e.g. the parts stand in this relation to the whole), others in the sense of the essence – the whole or the synthesis or the form. And again . . . or briefly the voluntary or involuntary agent sets going or arrest the transformation or movement.

And finally, there is the goal or end in view, which animates all the other determinants factors as the best they can attain to; . . .) (*Physics* 195a20 – 26).

According to the above quotation, it seems that the end and agent are the main items (although between these two factors end is prior because it contains all of the other three), for the agent with regard to the end wants to transform matter into form. Now we want to apply appropriately this image in *Physics* to *Politics*. It can be said that book 1 in its usual position in most editions is an introduction, but more precisely a social and prepolitical introduction. The treatise is on politics and logically the focus should be on the polis, but according to the structure of it there are some phases, the most important of which is the household, before the appearance of the polis that we call *prepolis*. Therefore, it is better that the discussion begins with the household, its parts and management. As Aristotle says:

> επει γαρ οικια μεν πασα μεροσ πολεωσ, ταυτα δ, οικιασ,
> την δε του μερουσ προσ την του ολου δει βλεπειν αρετην,
> (Every household is a part of a state, and these relationships are part of the household, and the excellence of the part must have regard to the whole.)[6] (1260b15).

It seems logical that discussion about the first and primary stage that is the prepolitical be the first book, and after that the political or polis stage be the second book. It is possible to introduce this new phase according to different standards and perspectives. Aristotle's choice is a normative one that points to the best and virtuous polis. But why? According the rationale of classical political philosophy this kind of thinking is necessarily normative, and it seems that he wants to mention and emphasize the normative side or aspect of the transition from the past prepolitical stage to the new polis one. As a natural human fact we experience the household, but we have to pass through this as if through a necessary but incomplete part. Transition through the household is obligatory and a duty, but a duty that can only be realized in the form of polis—the polis as it must be. Thus polis or the political stage must be introduced normatively as a complete whole toward which the household strives, and we can see this theme in the second book of the *Politics*.[7] But for Aristotle polis is not merely an objective and material fact, but it has a subjective and spiritual side in which citizens, rather than households, are its more basic parts. All of this can be found in the wording of the third book:

> επει δ, η πολισ των συγκειμενοων καθαπερ αλλο τι των ολων
> μεν συνεστωτων δ, εκ πολλων μοριων, δηλον οτι προτερον ο
> πολιτησ ζητητεοσ. η γαρ πολισ πολιτω–ν τι πληθοσ εστιν
> (But a state is a composite thing, in the same sense as any other of the things that are the wholes but consist of many parts; it is therefore clear that we must first inquire into the nature of a citizen; for a state is a collection of citizens) (1274b 40, 1275a).

For Aristotle the discussion about the subjective side of the polis, or its citizens, is prior to the different structures or constitutions that it will take. Accordingly, the polis as a specifically human entity has a spiritual component, citizenship in its universal sense. But the polis is affected by and is situated in different kinds of structures that have different formulations of citizenship as a general category. With regard to book 3 and its overall normative context, it seems that Aristotle is searching for and wants to pinpoint normative citizenship as the matter and the constitution as the form and their interconnections.

If Aristotle's introduction to the political or polis stage is normative (books 2 and 3 are interconnected with each other as they appear usually in the editions) they set the stage for the "actual" and "ideal" constitutions or polities that are discussed in the rest of his treatise. Although the order of these two different kinds of polities is important in itself, for we can apply Aristotle's discussion on the four causes (*aetiology*) in both actual and ideal parts, the aim of the present chapter is not because we concentrate only on the ideal polity.

In regard to the *Politics* we have said that book 1 is a social introduction to the political and normative stage in book 2. After these introductory books, it seems that Aristotle suggests something about the "matter" (in its technical Aristotelian meaning) of the polis. The discussion about the "Polis as it must be" in book 3 is interconnected with the book 2, and it suggests that the polis as a normative human entity should have special and definite matter. In this "special" human entity we first encounter human beings as human beings, but the adjective "special" points to the political aspect of the polis. Thus, human beings should be transformed to citizens, for being human is not sufficient, and this theme makes the nature and essence of citizenship crucial and vital. At the same time we should not forget that citizens live and practice in the context of different kinds of political structures as the "form" (in its technical and Aristotelian meaning.). Thus, in the book 3 we have the interrelation of both the "form and matter" of politics par excellence. Yet, although the constitution is the basis and forming factor of citizenship, we actually encounter different constitutions. According to the different forms, therefore, we will have different citizens as the spiritual matter of politics. Aristotle points to this problem in the articulation of this abstract principle:

> δει δε μη λανθανειν οτι των πραγματων εν οισ τα υποκειμενα διαφερει των ειδει, και το μεν αυτων εστι πρωτον το δε δευτερον το δ, εχομενον, η το παραπαν ουδ εϖεστιν, η τοιαυτα, το κοινον, η γλισχρωσ. (But it must not be forgotten that things in the case of

> which the things to which they are related differ in kind, one of them being primary, another one secondary and so on, either do not contain a common nature at all, as being they are, or barely do so.) (1275a 35).[8]

Aristotle's discussion about the "form" or constitution as a whole necessitates his attention about the "matter" that we begin politics with. Human beings are the subjective and nonmaterial matter of every human community. But they are not enough for the political community as a general and unqualified entity; therefore, they have to take the special nature and essence of each specific qualified political constitution. Each kind of constitution as "form" forms its appropriate citizen as spiritual "matter." But as we pointed out, the overall environment of discussion is normative and the interrelations of form and matter function as the guiding model and framework for the rest of the treatise. In other words, according to our reading of Aristotle's *Physics,* the interplay of "matter" and "form" provides the substratum on the basis of which we can provide the overall skeleton of the *Politics.* In this environment we think, order, arrange, and understand the parts of Aristotle's treatise as the building blocks. Although in the related literature there are various theories and hypothesis about the arrangement of the books of *Politics,* there is a rather stable consensus about the usual and customary ordering of the first three books as they are. It is in this context and also commitment to understanding Aristotle as he would have understood himself that we referred to the second book of the *Physics* with regard to its theme on form and matter. We have tried to show that by appropriate application of this to the third book of *Politics,* we can discern the foundation of "form" (constitutions) and "matter" (citizens) as the normative model that can be followed for the rest of this political treatise.

According to our hypothesis there is no difference for arranging and understanding the rest of treatise in terms of the "actual" or "ideal" polities, for in both of them it is possible to find and explain the interactions of the four causes with special focus on the "form" and "matter." In this context we read books 7 and 8 (as they normally appear in editions) together as the Aristotle's ideal polity. To understand this polity four necessary dimensions that are scattered throughout different essays or chapters must be distinguished. The first chapter of book 7 begins with the dimension first in order and priority; the end or "telos" of the ideal polity—the highest good and the happiest life. Thus, a connection is made between the best life and the ideal polity; the best life individually and collectively is the final purpose of the best polity.[9] The common good for both the individual and community are the goods of the soul or the virtuous life. The other two external goods are goods of body that complement the virtuous life. Although all three goods are necessary, there is a hierarchy.[10] Thus, according to Aristotle:

> περι πολιτειασ αριστησ τον μελλοντα ποιησασθαι την προσηκουσαν ζητησιν αν–ακη διορισασθαι πρωτον τισ αιρετωτατοσ βιοσ.... διο δει πρωτον ομολογεισθαι τισ ο πασιν ωσ ειπειν αιρετωτατοσ βιοσ, μετα δε τουτο ποτερον κοινη και χωρισ ο αυτοσ η ετερος .... ωσ αληθωσ γαρ προσ γε μιαν διαιρεσιν ουδεισ αμφισβητησειεν αν ωσ ου τριων ουσων μεριδων, των τε εκτοσ και των εν τω σωματι και των ε–ν τη Φυξη, παντα ταυτα υπαρχειν τοισ μακαριοισ χρη.... αλλα ταυτα μεν λεγομενα [ωσπερ] παντεσ αν συγχωρησειαϖ, διαφερονται δ, εν τω ποσω και ταισ υπεροχαις. (The student who is going to make a suitable investigation of the best form of constitution must necessarily decide first of all what is the most desirable mode of life. . . . Hence we must first agree what life is most desirable for almost all men, and after that whether the same life is most desirable both for the community and for the individual, or a different one. . . . For as regards at all events one classification of things good, putting them in three groups, external goods, goods of the soul and goods of the body, assuredly nobody would deny that the ideally happy are bound to possess all three . . . But although these are propositions which when uttered everybody would agree to, yet men differ about amount and degrees of value.) (1323a 15,20,25,35).

With regard to the above "telos" and end (final cause), it seems more natural that Aristotle pay attention to the matter and form of this ideal polity that reminds us of the tenor of the book 3. We can consider the second chapter of book 7 as the "matter" dimension, and accordingly there should be suitable material to be worked on for reaching to the ideal and best polity. For Aristotle this material that is both human and nonhuman should be possible and have proper qualitative and quantitative characters. "These materials are first, human beings, for to make the best regime is the same as to make human beings virtuous, and, second, territory, for human beings must have somewhere to live and something to live off. Both these materials exist in the right amount and be of the right sort."[11] This guidance necessitates that in the different chapters of book 7 Aristotle speak about the quantity and quality of human and nonhuman categories. He proceeds to do so in his discussions of population; inhabitants and their the natural endowment; and the social strata and structures on one hand; and the territory and the physical planning of the central city on the other hand. This material is not complete and should be transformed and improved in a proper way toward the telos or goal of the ideal polity. Thus, according to Aristotle:

> αρχη των λοιπων ειπειν ορωτον ποιας τινας δει τασ υποθεσεις ειναι περι τησ μελ–λουσης κατ ευχην συνεσταναι πολεως. ου γαρ

> οιον τε πολιτειαν γενεσθαι την αριστην ανευ συμμετρου χορηγιασ. διο δει πολλα προυποτεθεισθαι καθαπερ ευχο–μενους, ειναι μεντοι μηθεν πολιτων και χωρας. (The starting-point for the remainder of our subject is first to specify the nature of the conditions that are necessary in the case of the state that is to be constituted in the ideally best manner. For the best constitution cannot be realized without suitable equipment. We must therefore posit as granted in advance a number of as it were ideal conditions, although none of these must be actually impossible.
> I mean for instance in reference to number of citizens and territory. (1325b 35).

Next and interconnected with the matter is the essence or "form" that this "matter" should accept. The forms of the "material" and nonhuman dimensions of the best polity are specified and clearly stated under such topics as the size (that has to be moderate), topology and site of the territory (that has to be surveyable), and the desired natural endowment (that has to be mixed). But the more important and detailed discussion is about the "form" of the "human matter" that is symbolized by the category of culture or "education." On the one hand, according to the first chapter of book 7, the "end" of the ideal polity is a virtuous life or felicity for the individual and community at large. On the other hand, the population and inhabitants as the subjective and nonmaterial matter of this polity must take and accept the necessary and proper "form" or essence that is suitable for the best polity. Both of these need education and culture (not merely training) for their realization. Accordingly, education is necessary for achieving the ultimate goal and also making virtuous citizens. In other words, there are three ways for both of the mentioned tasks—natural endowment, habit, and rational principle—and it is in regard to the last two that education and the legislative art, or more precisely education approved by the legislative are, are required.[12] Thus, according to Aristotle:

> Περι δε της πολιτειας αυτης, εκ τινων και εκ ποιων δει συνεσταναι την μελλουσα–ν εσεσθαι πολιν μακαριαν και πολιτευσεσθαι καλωσ, λεκτεον. επει δε δυ εστιν εν οις γιγνεται το ευ πασι, τουτοιν δ, εστιν εν μεν εν τω τον σκοπον κεισθαι και το τελος των πραξεως οπθωσ, εϖ δε τας προς το τελος φεπουσας πραξεις ευρισκειν.... Τα τρια δε ταυτα εστι φυσις εθος λογος. και γαρ φυναι δει πρωτον οιον ανθρωπον αλλα μη των αλλων τι ζωων.... Την μεν τοινυν φυσιν οιους ειναι δει τουσ μ–ελλοντας ευχειπωτουσ εσεσθαι τω νομοθετη, διωριασμεθα προτερον, το δε λοιπ–ον ερηον ηδη παιδειας. (We must now discuss the constitution itself, and ask what and of what character should be the components of the state that is to have felicity and good government. There are two things in which the welfare of all men consists: one of these is the correct establishment of the aim and end

> of their actions, the other the ascertainment of the actions leading to that end. . . . But there are admittedly three things by which men are made good and virtuous, and these three things are nature, habit and reason. For to start with, one must be born with the nature of a human being . . . Now we have already defined the proper natural character of those who are to be amenable to the hand of the legislator; what now remains is the task of education.)(1331b 25, 1332a 40, 1332 b10).

Accordingly, education as the conclusive "form" has interconnections with the subjective or spiritual "matter" and should regard all the different parts of the soul (not be one-dimensional) and also of life (be comprehensive). This kind of form, or culture and education, will be implanted and institutionalized gradually in a two-stage order that consists of a preliminary education in infancy to age seven, and the advanced or proper education that has division within.[13]

Thus far we have identified three causes or factors: the "end" or purpose of the ideal polity, and the human and nonhuman materials which must be formed and take the proper nature or "essence" through education. But who is or are the agent(s) of this transformational process? It is the subject that is introduced in the book 8 under the title of the education of the youth (or majority, for usually and in more polities the youth are in the majority.) It seems that it is possible, with education, to transform the youth into the qualitative majority. This kind of educated majority as agent has two actual functions with regard to the construction of the polity and its doings within. According to Aristotle:

> Οτι μεν ουν τω νομοθετη μαλιστα πραγματευτεον περι την νεων παιδειαν, ουδει–σ αν αμφισβητησειεν. δει γαρ προσ εκαστην παιευεσθαι, το γαρ ηθοσ τησ πολιτε–ιασ εκαστησ το οικειον και φυλαττειν ειωθε την πολιτειαν και καθιστησιν εξ α–ρξησ. (Nobody would dispute that the education of the young requires the special attention of the lawgiver. . . . For education ought to be adapted to the particular form of constitution, since the particular character belonging to each constitution both guards the constitution generally and originally establishes it.) (1337a I).

An educated, youthful majority, it seems, will be the founder and maintainer of Aristotle's ideal polity. In ordering and understanding Aristotle's *Politics* we have used his etiology in the second book of the *Physics. Physics* suggests that the four causes can be compressed into the two categories of "form" and "matter," and then for more elaboration they can be discussed in their fourfold classification.

ENDNOTES

1. Aristotle, *Politics*, trans. Ernest Barker (Oxford: Oxford University Press, 1976), xxxvi.
2. Aristotle, *Politics*, xxxix.
3. See Kenneth A. Telford, "A Rereading of Aristotle's Politics," *Skepsis* 12 (2001): 110–22.
4. Aristotle, *Physics*, vol. IV, trans. Philip H. Wicksteed & Francis M. Cornford (Massachusetts: Harvard University Press, 1970), 103, 104.
5. Aristotle, *Physics*, 133.
6. All bilingual quotations are from Aristotle, *Politics*, trans. H. Rackham (Massachusetts: Harvard University Press, 1972).
7. Peter L. Philips Simpson, *A Philosophical Commentary on the Politics* (Chapel Hill: The University of North Carolina Press, 1998), 72.
8. Aristotle, *Politics*, 94.
9. Simpson, *A Philosophical Commentary*, 195, 197.
10. Aristotle, *Politics*, 279.
11. Simpson, *A Philosophical Commentary*, 210.
12. Aristotle, *Politics*, 300 (with some alterations).
13. Simpson, *A Philosophical Commentary*, 244, 205.

PART II

HEAVENLY PERFECTION AND PSYCHIC HARMONY

Chapter 5

Making "Men See Clearly": Physical Imperfection and Mathematical Order in Ptolemy's *Syntaxis*

Michael Weinman

In *Syntaxis,*[1] Ptolemy tells us that our inquiry into the unchanging nature of the uniform motion of the heavenly bodies through the zodiac will help regulate the natural disorder of our souls and will, "most of all things, concerning that which is most beautifully good with respect to actions and to character, make men see clearly."[2] Why can Ptolemy be so confident of this? He claims a causal connection between the mathematical study of the heavenly appearances and the state of a human's soul, finding that "from the contemplation of the divine beings such as this, in its good-orderliness, symmetry and calm, those being carried after are made lovers of this divine beauty, and are made accustomed to, and re-natured, as it were, to a similar spiritual state."[3] Such claims regarding a close tie between the love of wisdom, especially of the highest objects of knowing, and a well-ordered character are not uncommon. This one is of special interest because Ptolemy emphasizes that it is only as one undertakes the study of the heavens as a mathematician, and not as a theologian or a physicist, that one can so improve the state of one's soul. I will investigate here what exactly about the mathematic study of the heavenly motion underwrites this claim.[4]

Working carefully through what Ptolemy calls the "demonstration of the apparent anomaly of the sun,"[5] in contrast to the corresponding attempt (and failure, we will argue) with respect to the planets, we will find that he aims to offer not an account of the path (or orbit) of the sun, or even some such mathematically possible, if "unreal," orbit.[6] Mathematical understanding, we will see, means precisely not looking for such an orbit, but rather somehow seeing the uniformity in the holistic character of the motion, not as an unfolding movement from here to there and back again, but as one unchanging thing: the very being of order. It is in coming to understand this about the sun and its motion that Ptolemy means for his work to contribute to the ordering of our souls as students of mathematics, and not of physics or theology.[7] Having worked through this account, I will ask what it might mean to engage in such an activity today, if such is even possible.

§1. *Syntaxis* on the simplicity of the sun's motion and compound planetary motion

The core of what we must understand from *Syntaxis* is a methodological difference between (a) the analytic tools ("hypotheses") Ptolemy is able to deploy in "saving" the "true" regular and circular motion of the sun around the earth from the "apparent" irregularities and noncircularity of that motion and (b) the tools he deploys in attempting to similarly hypothesize regularity in the motions of the planets. What is essential in this difference, structurally, is that the tools deployed in the first instance are "simple," while those for the planets are not. This "simplicity" is, for Ptolemy, both a feature of the analytic work that a mathematician does in approaching the subject matter, and also of the subject matter itself. Thus, if a mathematical account fails to meet the test of simplicity, it means either that the subject matter itself is not simple—this is impossible in this case, as the motion *must* be regular and circular (hence simple)—or that the mathematical analysis has failed to convey its full truth.[8]

The crucial starting point of Ptolemy's discussion of the motion of the sun is the assertion that: "The anomalies of the heavenly bodies result from the placement and order of their circles in the spheres, through which they make their motions; thus, there is nothing in the disorder which they are supposed to display that is alien to the eternality of their natures."[9] The "reason for the appearance of this anomaly," he continues, can be demonstrated by the two "first and simplest" hypotheses: the eccentric and the epicyclic. We will briefly present these hypotheses and how they preserve the uniformity and circularity of the sun's motion; first, though, we ought to note that what we demonstrate, on either hypothesis, is not a feature of the motion itself, but rather the cause of there appearing to be something irregular about the motion. That is, we are not concerned with the measurement and explanation of a physical phenomenon; rather, we are demonstrating how it is possible that a heavenly body can "appear to our eyes to cut unequal arcs in equal times."[10] Having said that we can demonstrate the cause of the appearance of anomaly, it is given that we can and must accept the reality of uniformity.

For reasons of space and the scope of the current argument, I will not go through Ptolemy's demonstration of the hypotheses, or the proof of their application to the anomaly of the sun. Rather, I will present the statement of the ratios on the two hypotheses which makes possible the appearance of "unequal arcs in equal times" generally, and then describe how Ptolemy makes these ratios determinate with respect to the motion of the sun. This will suffice to make clear the difference between the demonstration with respect to the sun and with respect to the planets, which is our focus.

Ptolemy begins by describing each hypothesis, emphasizing that on either, there is an "angle of difference" which accounts for the appearance of irregularity in what is really a uniform circular motion: on the epicyclic hypothesis, we are to imagine that the heavenly body is in regular circular motion around a center which itself is in regular circular motion around the center of the world, which coincides with center of the earth (and which Ptolemy calls the "center of vision"); on the eccentric hypothesis, we are to imagine that the heavenly body is carried in regular circular motion around a center which is along the diameter of the "great circle," but which differs from the "center of vision," by a certain amount.[11] In the epicyclic hypothesis, the "angle of difference" is the angle at the center of vision, bounded by (a) the center of the epicycle and (b) the apparent position of the body; on the eccentric hypothesis, it is the angle at the apparent position of the body, bounded by (a) the center of the eccentric circle and (b) the center of the vision. The hypotheses turn out to be equivalent in describing the phenomena, owing to a sameness of ratio which itself arises from this same angle being preserved (in different ways) on each of the hypotheses.

Ptolemy explains that because "whenever (on the eccentric hypothesis) the ratio of the eccentricity [the distance from the center of the eccentric circle to the center of vision] to the radius of the eccentric circle is the same as (on the epicyclic hypothesis) the ratio of the radius of epicycle [the distance from the body to the center of its motion] to the distance from the center of the world to the center of the epicycle," the same phenomena "are indifferently described by either of the hypotheses."[12] Indeed, he says, that for such phenomena—that is, all phenomena which display just one anomaly—"one single hypothesis suffices." That is to say, it is a matter of indifference to Ptolemy whether we "imagine" the body on an epicycle or an eccentric circle: it really doesn't matter what description or model of the apparent motion we provide; the truth is found in learning enough about what we see to forget about the visible altogether.

This is remarkable. What Ptolemy is asserting here is that, as mathematicians, we need not be concerned with finding an account that accurately describes the "actual phenomena," or even one that models them convincingly. This makes perfect sense, however much it differs from our intuitions: for Ptolemy, the appearance (specifically, the appearance of irregularity) is not to be explained; rather, it is to be explained *away*. At the same time, Ptolemy hints that with respect to those bodies which display a "double anomaly," it will be possible (and necessary) to "combine the hypotheses";

literally, to apply them one atop the other.[13] He is here pointing toward the presentation of the anomaly of the planets, and already telling us how such a demonstration will crucially differ from the demonstration for the sun.[14] For now, let's complete our discussion of the latter, and see just how Ptolemy succeeds in his project of a purely mathematical study of the "true" order which pervades the "appearance" of disorder. We already see, though, the seeds of the corresponding failure with respect to the planets to which we will have to attend momentarily.

What Ptolemy next needs to show is that those ratios are, in fact, the same because they result from one "angle of difference," which is the source of the appearance of irregularity, whichever of the hypotheses one adopts. This is achieved[15] by demonstrating that, selecting a position of the body at random, the same anomalistic difference (the amount by the body appears to have fallen short of or exceeded its true motion over a period of time [measured by the arc of the "great circle"]) is given on either hypothesis. While it is beyond our purview to carefully trace steps of this demonstration here, we can note that it relies on these steps: (1) fixing the positions of: (a) the body; (b) the center of the epicycle; (c) the center of the eccentric circle; (d) the center of the world; (2) showing that the line connecting (a) and (b) is parallel to that connecting (c) and (d), and thus, so too are the lines connecting (a) and (c) and (b) and (d); (3) inferring, there from, that the arcs subtending lines (a)(b), (b)(d) and lines (c)(d), (a)(c) are similar; and thus (4) that the body "ought" to be carried in equal times along these arcs; and thus (5) that the irregularity that appears at this randomly chosen position is given by the "angle of difference" as defined above.

The last step Ptolemy needs to make in fully accounting for (and thus, rendering irrelevant) the appearance of anomaly is to give us a measurement of just what this "angle of difference" is for the motion of the sun. Again, we will not here recreate his demonstration in its entirety; instead, we aim just to recapitulate those elements which are responsible for both the success in demonstrating the solar anomaly, and the failure with respect to the planetary anomalies.

Ptolemy begins by placing the sun's eccentric circle—crucially with a "chance radius"[16]—within the "great circle" of heavenly motion (the ecliptic). We stress this "chance radius" because this shows that the actual distance between the sun and the earth is irrelevant for Ptolemy here.[17] To be sure, if he were concerned with describing the actual appearance of the sun in its motions, or even with providing a model of same, he would not be so indifferent. Further, as we will see, he is forced to describe distances in accounting for the planetary anomaly. The only thing that is of importance to Ptolemy is that the radius of eccentricity (it could just as well be the epicyclic radius, as we will momentarily see) bears to the distance between the center of the sun's eccentric circle and the center of vision the ratio of 24:1. And this radius is important only insofar as it provides the means by which to compute the angle of anomaly at the quadrant (that is, at the position of greatest apparent difference between the true and apparent motion, which Ptolemy calls the "greatest anomalistic difference."[18])

Having established this ratio of 24:1, he simply draws the diameter of the eccentric circle, and then describes that circle, and having connected the center of the eccentric circle to the position of the body, then connects the center of vision (itself having a distance from the center of the eccentric circle that bears the ratio of 1:24 to the diameter's length) to the position of the body, at a right angle. The angle at the body, bounded by the center of the eccentric circle and the center of the vision—which is the "equation of the anomaly,"[19] meaning that it describes the greatest anomalistic difference between the uniform motion and the apparently irregularity—is thus given as 2;23° (where 4 right angles make 360°). Ptolemy then does the same for the epicyclic hypothesis: he positions the body on the epicycle such that the angle opposite the right angle formed at the position of the body, bounded by the center of the epicycle and the center of vision—which is the angle at the center of vision, bounded by the center of the epicycle and the apparent position of the body and is the equation of the anomaly, only on the epicyclic rather than eccentric hypothesis—is equal to 2;23° (where 4 right angles make 360°).

With this demonstration of both the greatest anomalistic difference (which occurs at the quadrant—that is, at a right angle) and the fact that the same greatest anomalistic difference is found on either hypothesis, Ptolemy is able to give[20] an account of the expected apparent anomaly for each position of the sun along the arc of its circular motion, and therefore to present its "mean motion" which accords with its true uniform motion, and also to describe for any given day of measurement the expected apparent and true positions of the sun in the zodiac. He is also able[21] to account for the inequality of solar days, and the accrual of inequality for periods of years. All of this data, it is crucial to note, is presented as evidence of the unreality of the appearance, not of greater precision in describing the appearance. For Ptolemy, then, this accounting is meant to be a sign not of knowing the appearance better, but of knowing better than the appearance.

While this recitation of Ptolemy's demonstration of the anomaly of the sun cannot have hoped to have *proved* anything, we have seen why either hypothesis equally well serves to show us why it is that, if we hold the uniform circular motion of the sun around the earth as axiom, we can nevertheless rationally explain the precise nature of the deviation from such regularity that appears to the best observation of the appearance of that motion. This is precisely what cannot be said for his attempt to demonstrate the same with respect to the planets, to which we turn now.

Ptolemy begins the account of the motions of the planets with an explicit conjunction of his work here with the earlier demonstration of the sun's motion, asserting that we are to "demonstrate for the five planets, as for the sun and moon, that all the apparent anomalies are represented by uniform circular motions," which are "proper to their divinity," while any "disorder or irregularity" is "alien to their nature."[22] Such a demonstration, he insists, echoing the "philosophical preface" to the *Syntaxis* to which we attended at the outset, is "truly the proper end of mathematics as theoretical philosophy."[23] Immediately though, we run into a series of rejoinders, concessions, and apologies, which begin with difficulties owing to the paucity of data and problems with its arrangement, and growing to problems with our ability to determine planetary positions at all, and then rising to lack of "right angles with the great circle of the heavens," which was so crucial in determining the true motion of the sun.[24]

This summary of difficulties then culminates with a final apology that points directly to the nature of the planetary hypotheses themselves, and to the contrast between the simplicity of the account of the sun's true motion, and the necessarily compound character of any attempt at a similar account for the planets. Specifically, Ptolemy begs forbearance for three things: (1) using procedures not in strict accordance with theory; (2) making basic assumptions which arise not from first principles, but from a "long period of efforts"; (3) assuming a type of "motion or inclination of the circles" which is different for the different planets.[25] His justification is that it is only in this way that he was able to preserve uniform circular motion (to be precise: a motion compounded of a series of uniform, circular motions, but which itself is not uniform or simple) for each of the planets, and that when such motion is "held for all without exception, each of the phenomena are demonstrated with a more basic and universal principle than that of similarity of the hypotheses."[26]

For one sharing the basic assumption of this work, this is an understandable plea: the most inviolate principle is the perfection, order, and divinity of circular motion; anything else is to be abandoned for that principle. And, once we've accepted the other basic propositions of this text—like the existence of a heavenly sphere, the centricity of the earth, and the rest—then it is clear that we are going to have to violate some methodological principles (like similarity of the hypotheses, and like the indifference of the placement of the planes of the motions of the heavenly bodies) in order not to violate that most basic principle. Nevertheless, lurking within these three requests is what can only be called, in light of the "simplicity" test discussed above, the ultimate failure of the demonstration of the anomaly of the planets. (1) "Procedures" not in accord with theory (an example is the need to treat circles having a certain inclination within the heavenly spheres as though they were "bare" circles of the sort we saw in the treatment of the sun—which were expressly not presented as being of the heavenly spheres; indeed, said spheres themselves having been a matter of discussion only after the demonstration of the solar and lunar anomalies were complete), (2) the basic assumptions which are made because we know we need them (we can call these "formal heuristics"[27]), and (3) the lack of similarity between the hypothesis of the planets all point to one glaring fact: there is no simple hypothesis for the motion of each planet, or for planetary motion as a whole.

Ptolemy makes this confession when he introduces the basic tool (hypothesis) by which he will take up each planet in turn: what came to be called the "equant," for which Ptolemy provides not a name, but a functional description.[28] In order to understand the description he gives, we first need to see how the account of planetary motion follows from the basic hypothesis described above, and is then set to work "simply" with respect to the motion of the sun. Ptolemy tells us that in attempting to preserve uniform circular motion for the planets, we actually have two anomalies to demonstrate: one is the anomaly with respect to the sun (that is, the planet fails to maintain a regular position relative to the position of the sun), and the other is with respect to the great circle.[29] In describing these anomalies, Ptolemy further explains that we are compelled to assign to the first (synodic) anomaly the epicyclic hypothesis, and will therefore assign to the second (longitudinal) anomaly the eccentric hypothesis. Ptolemy refers here to the original presentation of the hypothesis, acknowledging that, with the planets and unlike with the sun, we have to abandon the indifference of the hypotheses; this is already a large concession. We can no longer say that the anomaly is an appearance eradicated equally well by one of two hypotheses; rather, we need to combine our two hypotheses in one particular way—and in a different particular way—with each planet, in order to recognize the deeper order. But this is not the heart of the problem.

For that we must turn to the "equant."[30] This further hypothesis is needed, Ptolemy states, because we find that merely hypothesizing, as we did in the case of the sun, that the plane in which we draw the eccentric circle is stationary, and that the straight line through the two centers [of the eccentric and of the great circle],[31] cannot "work so simply" with the planets. Here, we note, simplicity returns as a failed norm: it is not that we cannot simply assume this or that,[32] it is that our hypothesis as employed to demonstrate (to explain away) the anomaly itself fails to go forward simply. This is a crucial point: as we noted at the outset, if the account of the heavenly body is not simple, then this can only mean (given what we mean by the divine simplicity of the heavenly bodies) that either there is some fault in our demonstration, or else that the bodies are not the perfect, simple, and unchanging beings we insist they must be.

Further, and here is where the hypothesis of an "equant" is introduced, we cannot (even) hypothesize that "all" we need to do is imagine that "the eccentric circle on which the center of the epicycle is carried is the same as the eccentric circle with respect to the center of which the epicycle moves regularly, cutting off equal angles at equal times."[33] Instead, we have to imagine another ("deferent") eccentric circle, sharing neither the center of the great circle (responsible for the true, "regular" motion), nor the center of the eccentric circle (responsible for the longitudinal anomaly), which is the center of the true "regular" motion for the center of the epicycle (responsible for the synodic anomaly); while the epicycle is apparently carried around the original eccentric circle. What can provide for any measure of regularity here is the "equant"; in Ptolemy's words, the hypothesis of (for all the planets other than Mercury) the "point bisecting the line joining the eccentric center producing the [longitudinal] anomaly to the center of the great circle" or (for Mercury) "the point whose distance from the center of the circle about which it rotates is equal to the distance of the latter towards the apogee from the center of the eccentric circle producing the [longitudinal] anomaly."[34]

It is this "equant" which, in the course of the individual demonstrations of the anomaly (in truth, the two anomalies) for each planet, will allow for a calculation of the combined (longitudinal + synodic) and individual (just longitudinal, just synodic) anomaly for each planet. In this respect, there is a real similarity between the achievement for the planets and the achievement for the sun. Nevertheless, beneath this similarity, is a fundamental difference. That difference can be seen in the fact that it is not possible, as it very much is possible with respect to the epicycle or the eccentric circle, to present this hypothesis as arising from first principles, and thus as a "simple," or part of a "simple," demonstration of apparent disorder masking true order.

The content of this difference is this: whereas, with the motion of the sun, Ptolemy is able to account for any "apparent irregularity" with the hypothesis of either an epicyclic or an eccentric circle, any regularity in the motions of the planets can only be "saved" using a single "deferent-epicycle" hypothesis, which not only relies on combining an epicyclic and an eccentric circle ("putting" the epicycle "on" the circumference of the eccentric circle), but also needs to specify the precise position of that eccentric circle with respect to some point on which the regular motion "ought" to be centered, but from which the "actual" center of the eccentric circle differs a certain precise amount. What's more, the regularity to be found in any of the motions of the planets needs to be "saved" with an independent deferent, and (as it happens) the deferents are also in different planes, and differ from the "true" center along axes which fall in different dimensions. This means that in order to "hypothesize" a regular motion for any one of planets—and surely to do so for all of them—we must place the arcs of the circles they theoretically trace in definite relation in a certain physical place.

It should be clear from even this highly condensed account, that while the account of the sun (allowing Ptolemy his assumptions, and eschewing for the present the hard work of following his account in detail for ourselves) passes the "simplicity test" described above, this account of the planets fails. We will next explore why this matters.

§2. Mathematizing a world which both is and is not both physical and theological

If successful, our close, if condensed, examination of the manner in which Ptolemy attempts to "demonstrate the anomaly" of the sun's motion and that of the planets has taught us that while the former can be understood, on Ptolemy's terms, as a purely mathematical enterprise, focused only on the rational demonstration of a conceptually possible order, the latter cannot be so understood, insofar as it requires us to imagine the regularity of motion being demonstrated to be actually compounded in the heavens in a particular way. But of what matter this distinction, beyond the analysis of Ptolemy's "essentially inaccurate" system? In this section, I intend to show that this difference is one that truly makes a difference, because it is precisely in requiring us to exercise our imagination in such a way as to "place" the heavenly bodies whose regular motion is saved despite apparent irregularity that the account of the planets—crucially unlike the account of sun—edges Ptolemy's account toward a physical one, and thus dispels the promise of a purely mathematical astronomy, an astronomy indifferent to "actual" physical positioning of the bodies under theoretical study.[35]

Having seen in the previous section how the account of the anomaly of the planets differs from that of the sun, we can now answer why this difference points to something as pivotal as a possible resolution, or better dissolution, of the contemporary antinomy between science and theology. The answer rests in the meaning of the mathematical failure of the demonstration of the planetary anomaly. Since Ptolemy is forced to place the planetary hypothesis within the heavens physically, and to theorize an epicyclic motion around an eccentric deferent, and a deferent particular to the given the planet, it becomes impossible to conceive of his account as anything other than a story—however plausible—of what is "actually happening" with this particular heavenly body and its motion. If Ptolemy, or his generous reader, must accept this as the meaning of his mathematical work, then there is nothing to do but understand his planetary hypothesis as an alternative to the heliocentric one and to weigh its merits as an explanation of the physical processes at work in heavenly motions. And, if that's the conversation, then clearly there is no interest in Ptolemy's work except an historical interest in prior, inaccurate explanations. Put another way, if the demonstrations of Book IX, as presented, are the best picture of a "mathematical" approach to the planets, then there really is no alternative but to abandon this notion of astronomy for the more accurate and consistent physical astronomy that we know today, and let that astronomy contest with theological accounts of the heavens for the right to speak definitively regarding cosmology, or the meaning of the heavenly bodies for human understanding.

Our hope—not as defenders of Ptolemy or some such, but as theorists of a possible third, mathematical, way between theological and physical explanations of the heavens today—is that there is some way to mobilize the account of the anomaly of the sun given in Book III. This requires that we take that account seriously. How can we do so today, as we reject the Ptolemaic system as a physical explanation, and as we acknowledge Ptolemy's failure to provide—indeed, the impossibility for anyone to provide—the purely mathematical astronomy he promises in his preface? Even if we have been convinced that there is this crucial difference between the accounts of the solar and planetary anomalies, what impact can this have on our overall assessment of the meaning or value of a truly "mathematical" astronomy?

A full answer to this profound question can only emerge if and when a new form of mathematical realism takes hold—that is, when the objects of mathematical understanding are taken to be something other than a model of reality, or a symbolization of physical laws, or just formally self-referential and unreal in any meaningful sense.[36] The present reflection on the meaning of the account of the anomaly of the sun can only be seen as a case study on what that understanding might look like. That is, what is (for Ptolemy, in the account in the *Syntaxis*) true of the seeming disorder, and deeper order, of the motion of the sun—that being the mathematical object under consideration in this work—will be true of all mathematical objects when we engage in this broader project. Only as we see the fruit of this larger work will we truly be able to assess the viability of any "mathematical" enterprise. Why should such a project, radical in its nature and so seemingly hopeless in its ambitions, be taken seriously?

Because what is true of our mathematical object here is nothing less than the true meaning of order: order is not the periodicity of the "actual" motion we "see" the sun making; order is also not the abstraction of that motion into some orbit which exists as a model of the "true" perfect motion of the body that is hidden in the appearances; order is not the formula that gives us the quantitative measurement of that motion. Rather, order is the condition that exists in us, as we reflect on the wholeness of the motion of the sun. This order, to be sure, is not a feature of the universe—one need not hold that the sun never came to be or will never pass away, nor that the earth is the center of anything so as to see this order. Instead, this order—a mathematical object on the understanding we develop here—is a feature of our understanding itself, and as such it has an ontological status we need to understand better.[37]

Having thus made explicit that this exploration, as it concludes, is only a prolegomena, let us return to our original question so as to take heart with respect to what makes the work before us worth doing. We wondered, you will recall, how it is that Ptolemy could be so certain that if, and only if, one engages in the work of developing a mathematical account of the motion of the heavens—and not a theological or physical one—one will be good. This seems to be a statement about a necessary connection not only between ontology and ethics, but also between theory and practice: if one gets the account of being right, then one will get the account of being good right, and if one gets the account of being good right, one will get the being good right. What does this statement have to do with the necessity of developing a thoroughgoing mathematical realism today, and the possible resolution of the debate between theology and physical science over the nature and provenance of the universe?

What we can now see is that it is because (in *Syntaxis,* which means "ordering-together") Ptolemy understands order *(taxis)* to be a feature of the reflective intellect in which order dwells, and which dwells on that order. If this—and not some feature of the cosmos as a system of motion, or some series of phenomena—is the being of order, then it follows that it is impossible to properly understand this being and not be in accordance with it. That is, if order is the dwelling of a stable whole in the intellect, and if people successfully direct their intellect to an understanding of that stable wholeness, then it is impossible for them not to have understood how they need to pattern themselves on that stable wholeness. For, in order to have successfully understood that stable wholeness, they have had to demonstrate the anomaly of the motion of the sun, and in order to do so, they have had to carefully work through the observations and also attended to the hypothesis, and then to have proved how the hypothesis can explain the contradictions between the suppositions and the observations. This demonstration, in turn, is impossible, if one has not already dwelt with one's intellect upon that order which must underlie the deceiving appearances. Finally, individuals genuinely understood how they need to pattern themselves on the stable wholeness, it is impossible for them not to experience a stable wholeness in their soul. In this way, for these reasons, Ptolemy is so certain both of a necessary connection between ethics and ontology and between theory and practice, given his fundamental views of the cosmos. He is thus sure that mathematics, and mathematics alone, can "make men see clearly."

ENDNOTES

1. All references to the *Syntaxis* (in its Arabic name, *Almagest*) will be cited by reference to volume number and pagination in: J.L. Heiberg, ed., *Claudii Ptolemaei Opera quae exstant omnia, Vol I: Syntaxis Mathematica*, 2 vols. (Leipzig: Teubner, 1898, 1903). The translation is my responsibility, though I have relied on (if deviated not infrequently from): G.J. Toomer, *Ptolemy's Algamest* (Princeton, NJ: Princeton University Press, 1998). For the Greek, I consulted N. Halma, *De La Composition Mathématique de Claude Ptolemee* (Paris: Albert Blanchard, 1988).

2. *Syntaxis,* H I.7 [i.e., Heiberg, Vol. 1, page 7].

3. *Syntaxis,* H I.7.

4. An opening caveat: owing to limitations both material (chiefly, the space I have to work with here) and formal (chiefly, my aim in presenting Ptolemy's "demonstration of the anomaly" as a kind of case study for a new understanding of mathematical realism), I cannot hope either to present (a) an account of the arguments attended to here with reference to *Syntaxis* as a whole or (b) an account of the ethical dimensions of *Syntaxis* as a whole. For the former, an interested reader is directed to: Olaf Pedersen, *A Survey of the Almagest* (Odense: Odense University Press, 1974); for the latter, to: Liba Chaia Taub, *Ptolemy's Universe: The Natural Philosophical and Ethical Foundations of Ptolemy's Astronomy* (Chicago: Open Court Press, 1993). I have no quarrel with Pedersen in this respect; I am largely in agreement with Taub, but with some reservations, the central one being described below (see n. 7) in some detail.

5. *Syntaxis,* H I.216.

6. In this respect chiefly, among others, our work will vary greatly from the manner in which "Ptolemaic astronomy" is generally approached, where Ptolemy's "two-sphere system" is described in modern terms as an alternative to a Copernican system, with the actual nature of the physical universe taken for granted as the standard for judging the systems. See, for example, the account of Ptolemy's account of the anomaly of the sun in Thomas S. Kuhn, *The Copernican Revolution* (Cambridge, MA: Harvard University Press, 1957), especially 67–73.

7. It is in this insistence that my objections to Taub chiefly rest: she too sees mathematics as primary and self-justifying for Ptolemy, but takes this to mean that mathematics suffices in itself as the sort of practice others pursue as theology. It is thus not a surprise that when she takes up the sentence quoted in the opening paragraph (Taub, *Ptolemy's Universe,* 30ff.), it is an occasion to present her view that, for Ptolemy, mathematics is a pedagogical tool for those who wish to enter into a practice—in the Platonic tradition—which is, in her words "quasi-religious" (Taub, *Ptolemy's Universe,* 36). I will here try to develop the view that Ptolemy presents

mathematics as self-sufficing precisely insofar as the mathematical object (like the "order" he uncovers in the heavenly motions) is the true best object, without reference to any other divinity: be it the bodies themselves as divine, or else some immaterial divinity responsible for the bodies and their motion.

8. Ptolemy's view of simplicity as both a methodological principle and a feature of the cosmos can be found in *Syntaxis,* H I.201–3.

9. *Syntaxis,* H I.216.

10. *Syntaxis,* H I.217.

11. Compare with *Syntaxis,* H I.218.

12. *Syntaxis,* H I.219.

13. *Syntaxis,* H I.219.

14. It is thus disappointing that Kuhn's canonical treatment of astronomy's development, describes Ptolemy's accounts of the planets and the sun as alike using a deferent-epicycle hypothesis to explain the anomaly (Kuhn, *The Copernican Revolution*, 67–68). As this introductory proviso makes perfectly clear, this cannot be right.

15. In *Syntaxis,* H I.226–27.

16. *Syntaxis,* H I.234.

17. Ptolemy does, of course, go on to provide a demonstration of the distance between the earth and sun in the course of constructing of the astrolabe in *Syntaxis,* V (H I.422–25). My argument here is not to say that the position of the body is irrelevant, simply, but rather that it has no relevance (and would actually be a hindrance) to the successful demonstration of the anomaly. The fact that the same cannot be said with respect the planets shows that the planetary anomaly has a different ontological status, and thus must be dealt with using ontologically different tools.

18. *Syntaxis,* H I.238.

19. *Syntaxis,* H I.239.

20. In *Syntaxis,* H I.240–58.

21. In *Syntaxis,* H I.259–65.

22. *Syntaxis,* H II.208.

23. *Syntaxis,* H II.208.

24. Compare this sentence with *Syntaxis,* H II.208–11.

25. Compare the preceding and following sentences with *Syntaxis,* H II.212.

26. *Syntaxis,* H II.212.

27. As noted above, our objection to Kuhn, *The Copernican Revolution,* was that he conflated the account of the sun and that of the planets—here we can see why this is of such importance for the thesis here. We are saying that Ptolemy himself confesses that with respect to the "double anomaly" of the planets, he is jerry-rigging a hypothesis after the fact; he is light-years away from a similar confession about the sun. The history/philosophy of science and history/philosophy of mathematics conversation is rather tone deaf to this difference as a whole. For an account of how (with a helpful list of central participants in the error, and citation of a helpful outlier), see: Jody Azzouni, *Knowledge and Reference in Empirical Science* (New York and London: Routledge, 2000), 25–26n.

28. *Syntaxis,* H II.252.

29. Compare the preceding sentence and the following with *Syntaxis*, H II.251.

30. Compare this paragraph with *Syntaxis*, H II.252.

31. This is why, in order to maintain the symmetry achieved by "that line through the centers of the eccentric and great circles," Ptolemy will have to introduce the "line of apsides"—and a different such line for each of the planets. See *Syntaxis,* H II.257f.

32. Which Toomer takes to be equivalent, though he notes in a footnote (*Almagest*, 442n.41) that Ptolemy attributes of the failure not to the manner of our assumption, but to the working of the hypothesis.

33. *Syntaxis,* H II.252.

34. *Syntaxis,* H II.252.

35. If the account of the planets were the only story in *Syntaxis,* then it might not be unfair to call mathematical study a case in "formal heuristics," where we attempt to present the regularity in some physical phenomenon, without claiming any kind of "to the ground" epistemological or ontological justification. This is the way Ptolemy is often discussed (when at all) in the philosophy of science and philosophy of mathematics literature. For a clear presentation of how such a reading works—and a hint as to why it is unfair—see Azzouni, *Knowledge and Reference,* 24–26 (and n. 27 above).

36 To be clear, I am not speaking here of the kind of mathematical realism often called Platonism (that the mathematical forms are constitutive of, or even the cause of, physical reality)—see the critical discussion of same in John D. Barrow, *Pi in the Sky* (Oxford: Clarendon Press, 1992), 249–84—but rather a view of "real mathematics" which defines mathematical normativity independently from either any application of mathematical truth or reference to external verification. I aim to present Ptolemy's work in *Syntaxis* as a case study in understanding mathematics differently from the way it has generally been understood under the dominant theory of formalism. An excellent overview of formalism, its roots and its sway, can be found in Philip J. Davis and Reuben Hersh, *The Mathematical Experience* (Boston: Houghton Mifflin, 1981), 330–87. This "other" understanding of mathematics, the different form of mathematical realism for which this presentation of Ptolemy is meant to be a schematic prolegomena, aims to depict mathematics as an ontologically meaningful practice whose objects are both real and not external to the practice itself. Such an account can be found (for instance) in the following, for all their differences: Jody Azzouni, *Knowledge and Reference;* John D. Barrow, *Pi in the Sky,* 284–95; and David Corfield, *Towards a Philosophy of Real Mathematics* (Cambridge: Cambridge University Press, 2003).

37. This is vague. I aim here not to argue for this claim, but to argue for the reasons why we ought to take seriously the need to argue for this claim. Such arguments can be found in Corfield's *Real Mathematics*, 233–35 and chapters 1 and 9, and in Azzouni, *Knowledge and Reference* and *Metaphysical Myths, Mathematical Practice* (Cambridge: Cambridge University Press, 1994).

Chapter 6

Realism and Liberalism in the Naturalistic-Psychological Roots of Averroës's Critique of Plato's *Republic*

Ahmed El-Sayed Abdel Meguid

Though it has been translated from Hebrew and published twice: first by E.J. Rothensal[1] and then by Ralph Lerne,[2] very little attention has hitherto been paid to Averroës's *Commentary on Plato's Republic.* However, being the only extant treatise that reflects Averroës's political philosophy, there has been a revived interest in it in the Arab world. In this vein, and given that the original Arabic version of this text is still lost, two recent Arabic translations have been published in the last three decades. The first by Hasan Majī d al-'Ubaydī[3] was based on the English translation by Ralph Lerner, and the second[4] by 'Ahmad Sha'lān and edited by 'Ābid al-Jābrī was made directly from the Hebrew text. Both translations tried to emulate as much as possible Averroës's Arabic style in his extant works. However, Sha'lān's translation being based on the Hebrew text tends to be more accurate and generally closer to Averroës's style. Thus the present paper will rely on this translation and on the extensive editorial comments and introduction of 'Ābid al-Jābrī.

This paper is not primarily concerned with textual and technical questions like whether or not this work is a *Commentary* on or an *Epitome* of Plato's *Republic* in the strict Averroësian sense. Rather, it puts forward a fresh reading of this key Averroësian work. Inasmuch as Averroës claims in the introduction that psychology should be the proper foundation of a proper political science, this paper will argue that Averroës's *Middle Commentary* more than the famous *Long Commentary on the De Anima* best explains his main critique of Plato in this work. In this vein, the second of part of this introduction is devoted to a brief historical account of the composition of Averroës's *Middle Commentary on the De Anima* and his *Commentary on Plato's Republic* and why they ought to be read in parallelism with each other. Second, the paper briefly summarizes Averroës's theory of the material intellect in the *Middle Commentary on the De*

Anima and points out what it takes to be the key liberal and realistic dimensions in it. Finally, the paper investigates the naturalistic-psychological grounds for Averroës' critique of what he saw as Plato's dialectical claims in the *Republic* about necessity in political change in general and the relation among the five forms of state in particular. In this respect, the paper demonstrates that Averroës's theory of the *Middle Commentary on the De Anima* adequately explains the scientific grounds of his critique of Plato. Through this demonstration the paper accentuates the liberal and realistic dimensions of Averroës's construction of Aristotelian politics through his naturalistic critique of Plato's *Republic.*

AVERROËS'S POLITICAL CRISIS AND THE COMPOSITION OF THE *MIDDLE COMMENTARY ON THE DE ANIMA* AND THE *COMMENTARY ON PLATO'S REPUBLIC*

Both Alfred Ivry in his introduction to the Arabic edition of the *Middle Commentary on the De Anima* and 'Ahmad Sha'lān's introduction to his translation of the Hebrew text of Averroës's *Commentary on Plato's Republic,* agree that both works belong to the last period of Averroës's intellectual production around 586 and 589 A.H. Alfred Ivry argues that based on Averroës's references to the *Long Commentary* in the *Middle Commentary on the De Anima* it seems that the latter belongs to the last decade before Averroës was persecuted by Almohad Caliph al-Mansūr Abū Yusūf Ya 'qūb around 592 A.H. In his own right Sha'lān bases his claim that the *Commentary on Plato's Republic* was written in this late period on the fact that many of the shorter *Epitomes* of Averroës were written in the later period. Both Sha'lān and Ivry strongly oppose the rather dogmatic assumption that Averroës first wrote *Epitomes,* then *Summaries* and then *Grand Commentaries* on Aristotle's works, and that these levels of commentary reflect a necessary chronological order to which Averroës adhered through out his career. Based on these arguments along with 'Ābid al-Jābrī's argument in *al-Muthaqafūn fi al-Hadāra al-'Arabya,* I will adopt the hypothesis that both the *Middle Commentary on the De Anima* and the *Commentary on Plato's Republic* were composed in the later period of Averroës's intellectual activity from 579 to 589 A.H. The discussion in the coming sections, particularly in section 2, will further reinforce this hypothesis.

THE MATERIAL INTELLECT AND THE PLACE OF MAN IN AVERROËS'S *MIDDLE COMMENTARY ON THE DE ANIMA*

As Davidson explains, Averroës, unlike his three key Arab and Muslim philosophical predecessors (al-Kindī, al-Fārābī and Avicenna) was particularly interested in qualifying the status of Aristotle's controversial passive of material intellect—ὑλικός νους as Alexander described it or παθητικός νους as Aristotle described it in *De Anima* 3.5.[5] In this section I will discuss Averroës's conception of the material intellect in the *Middle Commentary on the De Anima.* I will start with a brief sketch of Averroës's characterization of the material intellect in the *Middle Commentary* and the way it contrasts with his view in the *Long Commentary.* Subsequently, I will turn to a brief analysis of the implications this conception has for Averroës's understanding of human nature and his ontological status.

BETWEEN THEMISTIUS AND ALEXANDER OF APHRODISIAS: AVERROËS COMBINED THEORY OF THE *MIDDLE COMMENTARY*

Averroës devoted the third section of Book III of the *Middle Commentary on the De Anima* to put forward his interpretation of Aristotle's conception of the intellect. He starts the section by reiterating two of Aristotle's key arguments about the nature of the intellect. First, the intellect unlike the senses cannot have any specific nature because it must be capable of receiving all forms and not only a specific form as is the case in sense perception.[6] Second, the intellect is not affected like perception. It might be considered affected inasmuch as it receives material forms; however, it is not affected as senses are when they receive perceptible forms, for the latter change upon receiving these forms whereas the intellect does not change when it thinks.[7] The reason is that change in the case of the senses assumes a change from one state to the other; however, the intellect is one with its thought. It does not have a particular nature since particularity comes from the unity of form and matter but the intellect, unlike the senses, has no matter. It is thus active while being affected and vice versa.

After delineating the nature of the material intellect in a negative way by demonstrating logically what it cannot be, Averroës turns to the complicated question of positively defining its ontological status. In this respect, he explains that there are generally two theories of interpreting Aristotle's material intellect. The first one, which is by Alexander of Aphrodisias, claims that the material intellect is a mere disposition in the human soul like the disposition of tablet to receive writing.[8] The second one, which is by Themistius, but which Averroës also attributes to all other key interpreters, argues that the material intellect is an incorporeal substance.[9] However, both of them are rather problematic. Alexander's theory problematically posits an incorporeal disposition in the human substance. But this means that he posits an incorporeal being which must be separate from all matter in a physical substratum. Similarly, Themistius and the other commentators posit a disposition or potentiality in an incorporeal substance; but potentiality is associated with material passivity. To avoid the paradoxical positions of both stances, Averroës suggests to combine both of them in a new theory of interpreting the human material intellect. In this vein Averroës argued that the material intellect is a combination of both an incorporeal intellect that conjoins with man and a disposition in man. In doing so Averroës tries to avoid the problem of Alexander's theory which posits an incorporeal disposition in the human substance and the problem of Themistius's theory which posits a disposition or potentiality in an actualized incorporeal substance. Averroës thus writes:

> We may thus conclude that the material intellect is a combination of a disposition in us and an intellect that conjoins with this disposition. Inasmuch as this intellect conjoins with such disposition it is a potential intellect *('aql musta'id)* and inasmuch as it is not it is an active intellect. This intellect is indeed the active intellect whose existence will be demonstrated below. For inasmuch as it is in contact with this disposition it has to be only potential since it cannot think itself but is capable of thinking material forms. But inasmuch as it is not in contact with this disposition it has to be an active intellect thinking itself and not thinking what is here; I mean it does not think material forms . . . The forgoing discussion has thus shown to you that there are two theories about the active intellect: the theory of the Alexander and the theory of others. It has also been shown that the truth, namely Aristotle's theory, consists in combining both of them in the way we said. Through this combination we avoid positing an incorporeal being (*jawhar*) that contains a disposition in its substance. Instead we posit that this incorporeal being has such disposition not due to its nature but due to its conjoining with the substance that has such disposition, viz. the human being[10]

Through this theory, as Davidson rightly explains, Averroës avoids the defects in both Themistius and Alexander's theses:

> The untoward implication of locating the human disposition for thought in an incorporeal being are thus removed by locating the disposition *essentially* within the human organism, and the untoward implications of the locating the human disposition for thought in the human organism are removed by locating the disposition for thought, in an *accidental* fashion, in the incorporeal active intellect.[11]

The human material intellect is thus a disposition inasmuch as it belongs to the physical human substratum. But it is an incorporeal and separable inasmuch as an incorporeal being conjoins with the human substance which has this disposition and engages in the activity of thinking. It is an argument very similar to the one Aristotle presents in his *Physics* about a man qua man and a man qua a doctor or in his vocational capacity as a doctor.

But Averroës does not limit his qualification of the material intellect to this characterization. Throughout his discussion of the nature of the material intellect Averroës asserts and reemphasizes, as he did in the *Epitome of the De Anima,* the dependency of the material intellect on imagination. As a matter of fact he goes as far as claiming, as Alfred Ivry notes in his introduction to the Arabic edition to the *Middle Commentary,* that the material intellect is "the disposition of the imaginative faculty to think."[12] The reason is that all of the faculties of the soul are purposively connected with each other in a harmonious hierarchy with the intellect at its top. Imagination is disposed to think rationally and thinking wants to abstract rational forms from the material forms handed to it by the imagination and the senses. Therefore, as Davidson correctly notes, the material intellect has both a transcendent and an immanent dimension. The transcendent dimension is represented by the view of the active intellect as an incorporeal substance that activates the disposition for thinking through the thinking process that ultimately allows man to break from the immanence of spatial-temporal experience. By contrast, the immanent dimension is represented by the intellect qua an accidental disposition inhering in the incorporeal intellect.

If the activity of the intellect, being the highest faculty of the human soul, represents the highest form of the actualization of the human being as Aristotle claims in the *De Anima,* the *Nicomachean Ethics* and elsewhere, then the conception of the human material intellect is essential for understanding Averroës's conception of human nature. Accordingly, the second part of this section will put forward a possible interpretation of the implications the above view of the material intellect may have for Averroës's understanding of human nature in general and its ethical and political dimensions in particular.

THE MIDDLE COMMENTARY THEORY OF THE MATERIAL INTELLECT AND HUMAN NATURE IN AVERROËS

Unlike the *Long Commentary on the De Anima* where Averroës defines the material intellect to be a transcendent incorporeal substance in a way similar to Themistius's position, the conception of the material intellect presented in the *Middle Commentary* reflects a more individualistic perspective. As indicated above, the status of the intellect for the Aristotelian system reflects the essence of human nature in general. In the *Middle Commentary* the human intellect essentially connects the transcendent rational forms and the immanent empirical forms furnished by the imagination. This happens in two different modes. The human material intellect may abstract the rational forms from the material forms provided by the imagination which in turn depends on the senses. Or, the human intellect may be completely transcendent to all the immanent elements deduced from experience inasmuch as it thinks the first principles that are the basis of absolute demonstration (*burhān* or αποδείξει) and so the ground of ontology proper. The principle according to which this connection is made possible is, as Alfred Ivry explains, the teleological relation between the different powers of the soul.[13] Since the soul seeks a state of pure actuality and this state is fully achieved through the actualization of the highest power of the soul, viz. the intellect, each power of the soul as much as it possibly can desires to be rational. In this respect, since imagination, according to Averroës, furnishes the intellect with the material forms from which it deduces the rational forms, imagination is the power of the soul that is most disposed toward rationality. Similarly, the intellect, inasmuch as it thinks the first principles, desires to interpret and organize the intelligible forms abstracted from the imagination according to these principles. Thus the teleological structure of the relation between the different powers of the soul functions in two similar yet opposite manners inasmuch as it allows the material intellect to connect the transcendent with the immanent. On the one hand it sets the orientation of the imagination toward being rational and initiates its desire to make abstract the intelligible forms from the material forms it handles. On the other hand, it allows the intellect to interpret the first principles or more precisely apply them to the intelligible forms abstracted from the material forms to construct rationally coherent thought. The importance of establishing a connection between the transcendentally rational with the empirically immanent is therefore at the core of Averroës's theory of the material intellect in the *Middle Commentary.*

But what implications does this theory have for Averroës's conception of human nature in general? If the thesis proposed above is correct and the status of the human material intellect reflects Averroës's characterization of human nature in general then the following conclusions can be drawn. First, this conception of the human intellect is more in line with Averroës's view in *Long Commentary on the Metaphysics* and *Tahāfut al-Tahāfut* (*The Incoherence of the Incoherence*). As Davidson has explained, Averroës moved from his early emanationist account of the relation between the First Cause and the celestial spheres—especially the sublunar realm—in which he was more in line with al-Fārābī and Avicenna to a more naturalistic account.[14] The key reason underlying this transformation consists in his argument against the claim that from one cause only one effect can follow. Davidson correctly demonstrated that Averroës argued that this principle applies only to efficient causes but not for final causes. The intellect or the unmoved mover associated with every sphere thinks itself and thinks the First Cause; however, each does so in a different modality. Due to these different modalities, there are variations in the ontological status of beings in each and every sphere. If this interpretation is correct, this supports both Alfred Ivry's and 'Ābid al-Jābrī's theses, mentioned above, about the date of composition of the *Middle Commentary on the De Anima* in the later period of his life in which he composed his *Long Commentary on the Metaphysics.* The reason is that this view of the final cause can analogously justify Averroës's theory of the material intellect in the *Middle Commentary* and indicate the individualistic dimension in it. Inasmuch as each unmoved mover or intellect associated with each celestial sphere can think the First Cause in a different mode, so the active intellect once it conjoins with a human substance is actualized in a different modality. This modality is precisely the modality in which the transcendent rational forms are connected, related, and used to interpret the intelligible forms abstracted from the immanent material forms furnished to it by the imagination. This means that each and every human material intellect, inasmuch as it is actualized through thinking, actualizes a unique human being or substance; it is essentially connected with the imaginative faculty which in turn depends on sense-perceptual experience which varies in its scale and scope. Every human substance is actualized through the actualization of the human material intellect inasmuch as the latter connects the transcendent and the immanent dimension that is the function of experience.

The foregoing reading of Averroës's theory of the material intellect in the *Middle Commentary* also reflects a liberal and realistic tendency as much as it reflects an individualistic one. Based on the experience of each and every individual soul, the active intellect strives to actualize itself inasmuch as it is can fully connect what is rational and transcendent to what is immanent and empirical. This in turn implies that the ethical and political perfection of human nature is always connected with the reality and context wherein the human being exists, which is supported by Aristotle's arguments in the *Nicomachean Ethics.* It is thus an essentially realistic view. Further, the fact that each soul is in constant strife to actualize itself by bringing together the immanent and the transcendent dimensions leaves more latitude for the role human will may play in achieving human perfection.

Averroës's theory of the material intellect in the *Middle Commentary* is more in line with his later view on metaphysics in general. Second and more importantly its implication for Averroës's understanding and conception of human nature reflects a better appreciation of individualism and so variation among humans. This individualistic outlook has two key dimensions: a liberal and a realistic one. In turn, these two dimensions have important implications for Averroës's view of the ethical and political life inasmuch as the latter, in the Aristotelian system, primarily depend on the conception and the nature of the soul and its highest power, namely, the intellect.

AVERROËS'S CRITIQUE OF PLATO'S *REPUBLIC*

Why must psychology be the basis of political science?

Averroës starts the First Treatise of his commentary on Plato's *Republic* with an explicit statement of his intention:

> The intention of this is to abstract such scientific arguments attributable to Plato as are contained in the *Republic* by eliminating the dialectical arguments from it.[15]

In Averroës's language scientific arguments means arguments proven by demonstration from first principles, whether they are ones that he will demonstrate or ones already demonstrated in other sciences. Averroës then devotes the rest of the First Treatise to a discussion of the key points related to political science in general and to the *Republic* in particular including the necessary attributes of the ruler of the republic, the nature of the perfect state and the necessity of sharing women and children.

In the Second Treatise, Averroës demonstrates that the ruler of the perfect state must be a philosopher. Subsequently he argues that the ruler of the perfect state must be a philosopher primarily because he actualizes the end of humanity in the best possible way inasmuch as he attains the state of contemplation or thought thinking itself. Having demonstrated this point, Averroës then reinforces it by putting forward the following crucial part of his commentary on the *Republic:*

> We say that he who possesses this science joins with the physicist in inquiry concerning this object of search. For physics considers this end in so far as it is a perfection for it . . . it has already been made clear in physics that man is composed of soul and body; and that the rank of the body relative to the soul is that of matter and the rank of the soul relative to it is that of form; and that matter is for the sake of form and the from is for the sake of the actions or passions resulting from [the form]. This being so man's perfection and end are to be found in the actions that necessarily result from it [the soul] . . . a man's end and happiness are attained only if those of his actions that are specific to him are realized by him in the utmost good and excellence. That is why it is said in the definition of happiness that it is an activity of the rational soul that is in accord with what is required by virtue.[16]

In this statement Averroës establishes the key fundamentals of his framework for a proper political science. First and foremost as 'Ābid al-Jābrī correctly notes, physics and particularly psychology furnishes the ground for ethics and political science. Second, inasmuch as psychology inquires into the nature of human being it defines its *telos* or end. In its own right psychology has demonstrated that the rational part of the soul is the highest power of the soul.

All other powers achieve their excellence inasmuch as they strive and desire to be rational. Accordingly, the excellence of man as a whole is achieved inasmuch as the rational power of the soul is actualized to the utmost. However, Averroës, following Aristotle, also asserts that the excellence of the rational activity presupposes virtue or practical perfection that in its own right is actualization of the nutritive and appetitive parts of the soul in accordance with reason. Upon the attainment of this practical perfection one may then proceed to intellectual perfection achieved through the activity of contemplation or thought thinking itself. This view asserts Averroës's view of the teleological bond between the different powers of the soul.

In light of the foregoing, psychology provides the scientific foundation for judging what may be demonstrable in political science and what can thus be properly included in it, and what may not be deemed demonstrable or scientific and must thereby be eliminated from the corpus of political science. Therefore, psychology will, even if Averroës does not explicitly state it, be his main tool for abstracting the demonstrable arguments in the *Republic* and eliminating the dialectical ones.

In order to better appreciate the key criticism Averroës addresses to Plato toward the end of his commentary on the *Republic* and the role I will argue his psychology plays in it, I would like to briefly consider his analysis of the role of the will and choice. In the course of his discussion of the relation between the theoretical and the practical sciences and virtues, Averroës writes the following:

> It has been made clear in physics that everything that has potentiality in its existence [can attain] its ultimate perfection only when it exists as complete actuality with no potentiality at all mixed with it. It is for us to bring into being and to perfect the perfection and actuality that are of this character—i.e. it comes to pass through choice and will since there is no sufficiency in nature to bring it about.[17]

As has been indicated, Averroës insists on the interrelatedness of the practical and the theoretical. As 'Ābid al-Jābrī rightly comments in a footnote on this part of the text, Averroës sees that the will is necessary for the attainment of moral virtues and practical perfection. In doing so the will sets the stage for the Active Intellect to fully actualize the material intellect inasmuch as it conjoins with the human substance. In this context, Averroës wages a fierce critique against Plato and the Neo-Platonic emanationist theory for claiming that theoretical knowledge exists eternally and that theoretical perfection only consists in the act of recalling such knowledge. In this context Averroës argues that this would deny the role of the human will to attain practical perfection. However, Averroës does not elaborate more on why he insists on leaving room for the will and why it is so important for him in connection with theoretical excellence, the ultimate of form of human activity and excellence.

In light of the reading put forward previously of Averroës's theory of the material intellect in the *Middle Commentary*, I propose the following interpretation of Averroës's insistence on the role of the will. It has been shown earlier that the kernel of Averroës's theory of the material intellect is its function in connecting the transcendent and the immanent inasmuch as it is both an inseparable substance but also has a disposition inasmuch as it conjoins with the human substance. The implication of this theory is, as explained previously, the way it indicates the importance of the individualistic dimension in Averroës's psychology. The perfection attained by every human being is essentially connected with its experience inasmuch as the material intellect has this hybrid nature and depends on the material forms furnished to it by the imaginative faculty. Bearing this view in mind, Averroës's argument for the importance and the centrality of the role of the will may be understood. Since rational activity is the highest activity of the human soul and in which its perfection consists, and since this activity depends or must be connected with the experience of every human soul, the will, inasmuch as it is a key presupposition for attaining practical perfection, is indispensable for attaining intellectual excellence. Every individual thus bears part of the responsibility for achieving a perfect state and society.

In the next section I will show how the earlier suggestion of the connection between psychology and political science may explain Averroës's concluding critique of Plato's main dialectical false claim in the *Republic*.

THE TWO OBJECTS OF AVERROËS'S CRITIQUE: NECESSITY AND THE RELATION BETWEEN THE FIVE CONSTITUTIONS

After devoting a considerable part of his work to an analysis of Plato's theory of the five states or constitutions and the relation between them as explained by Plato in Books VIII and IX of the *Republic*, Averroës writes:

> One might object that if the case is as he [Plato] thought, namely, that there is among these cities what resembles two opposites—the two being the virtuous city and the tyrannical one—and that there is among them what resembles an intermediary between the extremes, then it does not necessarily follow that the change of cities proceeds according to this order. This exists only in natural things, for it is the way with to bring about opposites via intermediaries. As for these things, they are completely voluntary. And since all these natures—i.e.

> the nature we have described—are to be found in all these cities, it is possible for any city among them to change into any other. We say that what Plato said undoubtedly is not necessary but is this way as a rule[18]

Here Averroës argues that since the five natures of the human soul, which correspond to the five kinds of states or constitutions, are always present in any state, the transformation from one form to the other—whether to the better or to the worse—must not necessarily happen in the order described by Plato. Necessity is only found in natural things. There is no necessity in human affairs since they are absolutely voluntary and based on the will. But how is it possible that a devout Peripatetic like Averroës so argues? Is not the human substance part of nature and any phenomena pertaining to the human must be understood in terms of necessary causal relations? While a number of commentators including 'Abid al-Jābrī have credited Averroës's liberal taking on Plato, none tried to investigate the scientific psychological ground of such a critique.

It has been argued above that the individualistic tendency in Averroës's theory of the material intellect in the *Middle Commentary* matches the turn his view on metaphysics undergoes from an early emanationist outlook to a more naturalistic one. The basis of the analogy between the *De Anima* and the *Long Commentary on the Metaphysics* was based on the following rationale. Averroës breaks away from the emanationist outlook in metaphysics inasmuch as he argues that the argument that only one may proceed from one only applies to efficient and not final causes. Since the intellect associated with every celestial sphere thinks the First Cause in a different way this allows for different forms of souls and forms of life. Similarly, every human substance inasmuch as its highest faculty, namely the material intellect, is a hybrid entity that connects the empirically immanent with the transcendentally rational, will be actualized in a different manner. Further, since the will of every individual human substance plays a central role in achieving moral excellence, a necessary prerequisite for achieving intellectual excellence, this means that every individual in the society inasmuch as he strives to achieve perfection can attain to the best form of human perfection in his own context. This has important implications for Averroës's political theory. Inasmuch as individuals in the same political context have similar political experiences, any one of them who strives to achieve human excellence may lead the rest of the community to a better political state. This is why Averroës argues that Logic, the study of first principles of thought, ought to be the first thing taught. The reason is that every human being can then make sense of the path to perfection based on his own context. Since this context is shared by all members of the society they may, if they are all well-grounded scientifically, arrive at similar conclusions regarding the best way of attaining their perfection.

In this respect Averroës's psychology allows him to leave more room for the collective will of the people and its role in changing the political reality in the society. Further, Averroës's theory allows him to scientifically establish that every individual can change inasmuch as he strives to actualize his humanity in its fullness. Further, if education starts by illuminating the first principles and the rules of logic and demonstration, even if only few in the society seek to achieve, they can better lead the rest of the members of the society to a better state and social order for the following reasons. First, inasmuch as the actualization of human excellence consists in using the rational transcendent ideas to interpret the immanence of experience, and since every one has a solid background in logic and demonstration, those who have the will to a better society may more easily lead the society and convince it of the right political state. Thus, Averroës approaches modern political liberalism in emphasizing the role of having a basis for communication in the society in order to achieve political change. Second, Averroës seems to be less caught in the biological determinism to which Plato's theory of the metals lends itself. Inasmuch as everyone's intellect is of this hybrid nature and as long as everyone has a background in proper thinking then change is possible; even if this education is not available in corrupt states the very nature of the human material intellect allows for the possibility of change and guidance. No one is bound to a biological nature into which he was born and to which he is bound throughout life, whether bronze, silver or gold. Third, Averroës's theory makes room for the possibility of achieving the perfect state. Instead of being utterly pessimistic about the possibility of ever achieving a perfect state as Plato was due to the idealism of the theory of the forms, Averroës leaves room for achieving perfection. However, perfection for Averroës is not an idealistic moment like Plato; rather perfection is always bound to the human condition inasmuch as it represents the actualization and the excellence of the intellect, which in turn is always connected to the individual experience of this or that human substance. Averroës's notion of perfection is defined in terms of a realistic conception of the essence of human nature as partly transcendent and rational and partly empirical and immanent. This lets him allow for a more expanded role of human endeavor and strife in politics and changing the social reality.

Nonetheless, due to his realistic attitude Averroës realizes that Plato's assumption of the order the five forms of the state and the way they may transform into one another is likely. In this vein he writes:

> Since the *nomoi* and especially in the virtuous city, are not suddenly transformed—and this too is because of the habits and excellent states to which they accustom its members and with which they grow up—but rather are only transformed gradually and to what is proximate.[19]

Yet, Plato's account about necessity in politics and political change is only likely and dialectical but not demonstrable. Demonstration in political science should be based on its scientific foundation in psychology and such foundation, as has just been shown, indicates that there is no necessity in ethics or politics.

The suggested historical coincidence of the composition of Averroës's *Middle Commentary on the De Anima* and his *Epitome* or *Commentary on Plato's Republic* has led to two important results. First, it has been shown that Averroës's theory of the material intellect in the *Middle Commentary on the De Anima,* which reflects his general understanding of human nature inasmuch as the material intellect represents the status of highest power of the soul in potentiality, involves two important dimensions. First, it shows that the essence of human nature is partly rationally transcendent and partly empirically immanent. In this vein the excellence of the human being or human perfection consists in the capacity of each and every individual to anchor the rational ideas into the empirical and guide the empirical by the rational. Second, Averroës's theory reveals a significant appreciation of individual differences that is both realistically sensitive and more liberal. Second, it has been shown that the suggested historical parallelism between the composition of Averroës's *Middle Commentary on the De Anima* and his *Epitome* or *Commentary on Plato's Republic* is supported and helps illuminate a number of key arguments in Averroës's political theory as presented in the latter work. In this vein Averroës's hybrid theory of the material intellect helped illuminate the reason underlying Averroës's critique of Plato's claims about necessity in political change. Further, it reflected how realistic and liberal Averroës's conception of the perfect state and the possibility of its achievement are compared to Plato's.

The fact that Averroës did not have access to Aristotle's *Politics* allowed him to present his own Peripatetic political theory inasmuch as he tried to use his own readings of Aristotle's Psychology and Ethics to extract a science of politics from Plato's *Republic.* While the final picture he proposes for a proper Peripatetic political science is quite different from Aristotle's *Politics,* it represents a unique liberal and realistic endeavor in medieval political thought.

ENDNOTES

1. *Averroës Commentary on Plato's Republic,* Trans. E.I.J. Rosenthal (Cambridge: Cambridge University Press, 1956).
2. *Averroës on Plato's Republic,* Trans. Ralph Lerner (Ithaca, NY: Cornell University Press, 1974).

3. Hasan Majīd al-'Ubaydī, *Ibn Rushd: Talkhīs al-Ssyāsa l-'Aflātūn* (Beirut, Lebanon: Dar al-Tali'a, 1988).

4. Ahmad Sha'lān, ed. Muhammad 'Ābid al-Jābrī, Al-*Darūrīfī Al-Ssyāsa: Mukhtasar Kitab Al-Ssyāsa l-Aflātūn* (Beirut, Lebanon: Center for Arab Unity Studies, 1998).

5. Herber Davidson, *Alfarabi, Avicenna and Averroes on Intellect* (New York: Oxford University Press, 1992), 258–59.

6. Ibn Rushd ed. Alfred Ivry, *Talkhīs Kitāb al-Nafs* (Cairo: Al-Hay'a al-Masrya al-'Āma l-al-Kitāb, 1994), 121.

7. Ibn Rushd ed. Alfred Ivry, *Talkhīs Kitāb al-Nafs,* 121–22.

8. Davidson, *Alfarabi, Avicenna and Averroes on Intellect,* 259–60. Also see: Badawī *'Abd al-Rahmān, Shurūh 'la 'Ristū mafquda fi al-Yūnānya,* 31–42.

9. Davidson, *Alfarabi, Avicenna and Averroes on Intellect,* 259-260. Also see: M.C. Lyons ed., *An Arabic Translation of Themistius Commentary on Aristoteles De Anima* (London: Bruno Cassirer, 1973) 169–213.

10. Ibn Rushd ed. Alfred Ivry, *Talkhīs Kitāb al-Nafs,* 124–25.

11. Davidson, *Alfarabi, Avicenna and Averroes on Intellect,* 281.

12. Ibn Rushd ed. Alfred Ivry, *Talkhīs Kitāb al-Nafs,* 9.

13. Ibn Rushd ed. Alfred Ivry, *Talkhīs Kitāb al-Nafs,* 9–10.

14. Davidson, *Alfarabi, Avicenna and Averroes on Intellect,* 245–54.

15. *Averroës on Plato's Republic,* Trans. Ralph Lerner, 3.

16. *Averroës on Plato's Republic,* Trans. Ralph Lerner, 82–84.

17. *Averroës on Plato's Republic,* Trans. Ralph Lerner, 93.

18. *Averroës on Plato's Republic,* Trans. Ralph Lerner, 143–44.

19. *Averroës on Plato's Republic,* Trans. Ralph Lerner, 144.

Part III

SKEPTICISM, MECHANISM, AND THE NEW POLITICS

Chapter 7

Skepticism, Science, and Politics in Montaigne's *Essays*

David Lewis Schaefer[1]

PHILOSOPHIC SKEPTICISM

The core claim of today's "postmodernists" is to have liberated themselves from the "metaphysical" assumptions about the knowability of truth that supposedly characterized the Enlightenment no less than classical philosophy. If Aristotle's supposedly "teleological" view of the universe is now regarded as a relic of more naïve times, so too is the modern doctrine of natural rights.

I shall contend here that the postmodern claim derives from a superficial reading of the greatest philosophers of both classical antiquity, notably Plato and Aristotle, and early modernity. My primary witness will be Michel de Montaigne, whose *Essays* (1580–1592) was one of the most important and influential texts of early modern philosophy, particularly as a source of the sort of skepticism subsequently elaborated by such thinkers as Locke and Descartes. Montaigne is often viewed as a pre-eminent skeptic who challenged the ostensibly dogmatic character of classical, premodern philosophy. He is thought to have grounded his position in the doctrines of a largely obscure ancient philosophic sect, the Pyrrhonians, who denied the possibility of any sort of knowledge.[2] However, this interpretation overlooks the subtlety of Montaigne's account of Pyrrhonism in his "Apology for Raymond Sebond" (*Essays* II.12), in which he suggests that *all* of the ancient philosophers were skeptics in the original, Socratic sense—being aware of the uncertainty of all human beliefs (II.12, 487 [375]).[3] Moreover, underneath a surface of mockery, Montaigne praises the philosophers as the greatest of human benefactors. The seeming dogmatism of Plato and Aristotle, he explains, was only a protective mask that they donned to avoid being punished for their heterodoxy and unbelief—and in order to have a morally salutary effect on the nonphilosophic multitude (II.12, 481 [371], 487–93 [375–80]). In his subsequent catalogues in II.12 of the philosophers' various and seemingly ludicrous doctrines about the gods and the human soul, he intimates his doubt that they were "speaking in earnest" when they discussed such things; they concealed their awareness of their ignorance to accommodate "public usage" and "so as not to frighten the children" (498 [384]; 527 [408]).[4]

The very challenge that Montaigne purports to offer in the "Apology" to the claims of non-Pyrrhonian philosophers to wisdom actually recapitulates the Socratic critique of the pre-Socratic philosophers' approach, a critique that is mirrored in Aristotle's *Nicomachean Ethics* as well. At one point, for instance, Montaigne borrows approvingly from Plato's *Theatetus* (173d–174c) Socrates's account of the philosopher as liable to the "reproach" a Milesian woman directed against Thales for "continually spending his time in contemplation of the heavenly vault" while disregarding what was "at his feet," to the point of being ignorant of his neighbor or even of himself (II.12, 519–20 [402])—hardly an apt portrayal of Socrates's own way of life.[5] Shortly thereafter, Montaigne observes that every science "has its presupposed principles," while denying that human reason can provide firm ground for those principles: "the end and the beginning of knowledge are equal in stupidity" (II.12, 522, 526 [404, 407]). Superficially, it might appear that the essayist is offering a challenge to Aristotle, who in the *Ethics* attributes knowledge of both the first principles of demonstration and the ultimate particulars to a faculty he labels "intelligence" *(nous)*, knowledge that must be more certain than the reasoning that follows from it. It is in that context that Aristotle defines wisdom *(sophia)* as a combination of *nous* and scientific knowledge, and reports its attribution to thinkers such as Thales and Anaxagoras who investigated the cosmos while supposedly lacking practical wisdom.[6] But in fact, this claim forms part of a dialectical argument that culminates in the elevation of Socratic knowledge of ignorance over such supposed wisdom—as well as a denial, echoed by Montaigne (I.25, 135 [99]) of the separability of genuine philosophic wisdom from prudence.[7] Montaigne indicates his awareness of Aristotle's own freedom from dogmatism both by his previous reference to the philosopher's contradictory assertions as constituting "Pyrrhonism in an affirmative form" and by his present remark that the Scholastics have made it "a religious matter" to debate any of the Stagirite's "ordinances": it is Scholastic Aristotelianism, which has turned philosophy into "an empty and fantastic name," not Aristotle's own teaching, that Montaigne challenges as antiphilosophic (I.26, 159 [118]; II.12, 487, 521 [376, 403]). That Montaigne was not a serious advocate of the Pyrrhonian form of skepticism—which reportedly led its founder to deny (for instance) the existence of a cart that was about to hit him, compelling his followers to save him—he indicates through subtle mockery elsewhere in the *Essays* (II.29, 683–4 [533]).[8] What more fundamentally distinguishes Montaigne from his predecessors among the classical political philosophers such as Plato and Aristotle, he suggests, is rather his openly proclaiming his skepticism—even while covering its political and antitheological implications with just enough pretense of submission to ecclesiastic authority to avoid being punished for his heterodoxy. This intent is highlighted by the emphasis in his chapter "On The Education of Children," which describes how a young gentleman of nonphilosophic capacity should be reared, on the need to encourage youth to question all accepted doctrines, with specific reference to conventional religious beliefs—thereby indicating, I believe, the effect that Montaigne intended the *Essays* as a whole, with its stress on the uncertainty of human beliefs, to have on its readers (I.26, 150 [111]).[9]

However, as Montaigne's periodic remarks about the distinction between men of truly philosophic capacity and most human beings (e.g., II.12, 481 [371]) indicate, his project of disseminating a popularized skepticism is not grounded in a belief that most people are capable of (or interested in) achieving philosophic understanding. Rather, anticipating Nietzsche's description of Christianity as "Platonism for the people"[10] by blaming the classical political philosophers' public stance for generating "edifying" myths that ultimately issued in the evils of theocracy, religious persecution, and civil war, Montaigne endeavors to replace that public teaching with a new one that will direct most people away from the quest for transcendence, instead making them fully "at home" in the terrestrial world with its earthly pleasures. This entails "debunking" religious claims more openly than his classical predecessors chose, or dared, to do. (I return to this theme in the third section of this essay.)

Montaigne's popularized form of skepticism does not issue in an outright relativism. Rather, from the fact of the inevitable diversity of human opinions about the "sovereign good" or ultimate goal of life, by contrast with our undeniable instinct of self-preservation and aversion to bodily pain (I.14, 55 [37]; I.53; II.8, 365 [279], II.12, 561 [435]) , he derives the formula for liberalism: "since philosophy has not been able to find a way to tranquility that is suitable to all, let everyone seek it individually" (II.16, 605–6 [471]).[11] His denial that life has an overarching positive goal will be echoed by Hobbes and Locke, who make it the ground of the doctrine of natural rights, according to which government has only the limited purpose of enabling human beings to enjoy the security of life, liberty, and the pursuit of happiness rather than seeking to impose a particular vision of human excellence on them.[12]

Despite Montaigne's rejection of the public posture of the classical political philosophers, there is nonetheless a deeper link between his own form of skepticism (as distinguished from his public doctrine) and that which characterized classical philosophy as a whole, but which sets him apart from the Pyrrhonian skeptics in particular. As Montaigne observes, the aim of the Pyrrhonians was a settled condition of tranquility, or "ataraxy," resulting from having liberated oneself from "the agitations we receive through the impression of the opinion and knowledge we think we have of things" (II.12, 483 [372]). Elsewhere in the "Apology," in apparent contradiction to his account of the endless philosophic disagreement about the supreme good, Montaigne similarly claims that there *is* "general agreement among all the philosophers of all sects that the sovereign good consists in tranquility of soul and body"—though they differ about the means to that goal (II.12, 468 (468 [360]). But it is difficult to reconcile this account of the purpose of life with Montaigne's depiction of himself as being engaged in a lifelong pursuit of understanding (e.g. III.3, 807 [629]). Although he disclaims seeking any "glory" in life other than "to have lived it tranquilly" (II.16, 605 [471], this denial of any grander project is rendered suspect both by his ironic preface "To the Reader," in which he ostensibly tries to turn readers away (but actually lures them on) by maintaining that the *Essays* has only a "domestic and private" goal, and by the acknowledgment in the concluding chapter that his concern to benefit "the little people" *[les petits]* may have been motivated by a desire for glory ("To the Reader," 9 [2]; III.13, 1079 [844]).[13]

The remark just quoted, along with Montaigne's forceful critique of existing laws, customs, and forms of government in such chapters as I.23, I.31, I.42–3, III.7, and III.9, gives evidence that the *Essays* embodies a *political* project, to which there was apparently no parallel in the writings of the Pyrrhonians. Moreover, while it is of the essence of philosophy to weigh all opinions open-mindedly, and to welcome challenges to one's preconceptions as an aid to one's pursuit of knowledge (III.8, 900–902 [704–5]), Montaigne's portrayal of the Pyrrhonians goes beyond this, representing them as gamesters who would eagerly maintain the opposite of any proposition that was put before them, just in order to "create doubt and suspension of judgment" (483 [372]). Far from embracing their teaching, Montaigne acknowledges that the Pyrrhonians' determined refusal to commit themselves to any opinion, even provisionally—including the opinion that we are necessarily ignorant—makes the doctrine "difficult to conceive," especially since its authors "represent it rather obscurely and diversely" (485 [374]).

As Montaigne depicts the Pyrrhonians, far from being thoroughgoing skeptics, they appear to have been dogmatically committed to a doctrine or at least an attitude of their own, which no observation or argument could ever shake. By contrast, the outlook and way of life that Montaigne represents as his own in much of the *Essays* is a *zetetic* skepticism, one that engages in endless inquiry for the sake of achieving greater understanding, rather than resting content with the awareness that one is ignorant (III.13, 1045 [817–18]). In this fundamental respect, as well as his political concerns, Montaigne is faithful to the philosophic way of life exemplified by Socrates, which far from aiming merely at tranquility, seeks to awaken his auditors from their slumber, on the premise that the unexamined life is not worth living (III.8, 903 [705]; Plato, *Apology of Socrates,* 38a).[14]

Of course Montaigne's dedication to the pursuit of learning would be obvious from the abundance of classical quotations that fill the *Essays* (which are not there merely for show, he emphasizes [II.10, 387 [296]), and from his accounts of his enjoyment of philosophic conversation and "association" with books in III.3 and III.8. But there is another important way in which Montaigne departs from the Pyrrhonian doctrine, even in the very passages where he appears to cast doubt on the possibility of knowledge: his periodic remarks holding out the prospect of infinite future scientific progress (see especially II.12, 543 [421]; III.13, 1045 [817]).

It is paradoxical that an author who stresses the uncertainty of our judgment and our incapacity to penetrate the essences of things (II.12, 545, 547–8 [422, 424–5]) should express such hopes. The explanation of this paradox is that Montaigne espouses a new doctrine about nature that came to underlie the modern scientific quest: the view that since we cannot assume that we have access to the "real" natures of things, but can know only their sensible properties, we are free to reconstruct the objects we perceive—both in thought and ultimately in practice—with a view to better understanding and therefore making use of nature's mode of operation (II.12, 545 [422]). The ultimate fruit of this enterprise is a new science of "medicine," broadly understood (II, xxxvii, 745 [581]), that will engender what Montaigne's successor Francis Bacon will call "the relief of man's estate."[15] By making earthly life more comfortable and secure, the modern scientific project that Montaigne encourages harmonizes with his political project of making the generality of human beings "at home" on the earth.

SKEPTICISM AND SCIENCE

Since Montaigne is rarely viewed as an advocate of modern natural science—despite the recognition of textual linkages between his work and that of Bacon and Descartes[16]—establishing this claim will require a closer examination of the passages in which he drops hints to this effect. In both the "Apology for Raymond Sebond" and "On the Resemblance between Fathers and Children" (II.37, the concluding chapter of Book II), he offers ambiguous remarks which at first sight express unqualified skepticism regarding the scientific project, but which upon closer reading tend to endorse it. Both discussions also offer insight into the relation between Montaigne's teaching about science and his moral and political teaching.

In the "Apology," Montaigne professes to be "tempt[ed]" and "move[d]" by the "good writings" of the ancients, finding "each one right in his turn, although they contradict each other." Referring, for instance, to the suggestion by certain ancient thinkers that rather than being fixed in relation to "the sky and stars," it is the earth that moves—in contradiction to what human beings had believed for thousands of years—Montaigne adds that "in our day Copernicus has grounded this doctrine so well that he uses it very systematically for all astronomical deductions." But he infers, skeptically, that it seems "we should not bother which" of the two views is correct, given the possibility that "a third opinion, a thousand years from now, will . . . overthrow the previous two." Hence, he concludes, we have "great occasion to distrust" any "new doctrine," considering that it may have replaced an opposite one, and may some day in turn be "thrown out" in favor of another view (II.12, 553 [429]).

The conclusion that Montaigne draws, however, is not purely skeptical. Instead of giving up the quest for learning about nature as fruitless, since it seems to be a source of endless disagreement, he remarks that when "pressed with a new argument," one should judge that "what I cannot satisfy, another will satisfy." Rather than "believe all likelihoods," anyone who "finds himself weak" should "refer to wiser men, from whom he received his teaching" (429 [554]). He does not rule out the possibility that "wiser men" may be able to resolve the contradiction among rival scientific doctrines—at least until a better one comes along.

From astronomy, Montaigne turns to medicine, remarking how, long after the birth of medicine, the "newcomer" Paracelsus (who had died in 1541) "is changing and overthrowing the whole order of ancient rules," holding that until now "it has been good for nothing but killing men." While readily accepting that conclusion, Montaigne judges that it would be unwise to put his own life to the "test" of Paracelsus's "new experience" (or "experiment"). As we shall see, this is not Montaigne's final answer with regard to the possibility of medical progress.

Finally, Montaigne alludes to physics, citing an unnamed man who professed "novelties" and "reforms" in that field who told him that "all the ancients had evidently been mistaken about the nature and movements of the winds." When Montaigne asked him how the ancients nonetheless managed to navigate the oceans successfully, the physicist attributed it to "fortune." To this the essayist responded that he "would rather follow facts *[effects]* than reason," noting that these are things "that often clash"—as in a friend's report of seeing two lines meet after having "proved" that they could not do so (554–5 [430]).

Here again, Montaigne conveys a superficial impression of skepticism about scientific advance, only to correct that impression in the immediate sequel. Instead of endorsing the Pyrrhonian position, he remarks that the Pyrrhonians "use their arguments and their reason only to ruin the appearance of experience," taking advantage of "the suppleness of our reason" simply "to combat the evidence of the facts." Hence they purport to "demonstrate that we do not move, that we do not speak, that there is no weight or heat, with the same force of arguments with which we prove more likely things" (555 [430]).

Here, Montaigne clearly sets himself apart from the Pyrrhonians. Rather than denying the evidence of the senses, or that some things are more probable than others, the essayist chooses to be guided by that evidence, or to adopt a genuinely experiential or empirical perspective toward the universe. This theme reappears in "Of the Resemblance of Fathers to Children," in the midst of a seeming diatribe against medical science. Here he challenges physicians who cite "their experience as a warrant of the infallibility of their drugs," asserting that the greater part of medicinal powers consists of the "occult property" of simple substances, "concerning which we can have no other instruction than use." Among such proofs, Montaigne jestingly purports to accept those that reportedly derive from demonic inspiration, since he "never touch[es]" on "miracles." More seriously, he accepts those "proofs that are drawn from things that often fall into our use" in some other way—that is, effects that are found associated with some substance in everyday use. But he rejects as incredible "most of the other experiments *[experiences]* to which they say they were led by fortune or by chance"[17] (762 [594]).

What follows is an account of the conditions that an experiment would have to meet in order to generate medically reliable knowledge. Montaigne first cites the difficulties that obstruct such an enterprise: the need to isolate, out of the "infinite number" of substances, maladies, and circumstances the precise cause that happened to cure a given disease. Success in this enterprise would require that the investigator be guided by a "fortune" that is "perfectly artificial, regulated, and methodical"—rather than by a merely random process. Even if an experiment seemed to have demonstrated that a cure was effected by a particular substance, rather than by a host of other variables, the test would have to be repeated a number of times in order for a reliable "rule" to be derived from it (762–3 [594–5]).

Montaigne does not finally conclude that the foregoing difficulties foreclose all hope for progress in medical science. Rather, he laments that out of the millions of human beings who have lived, only three physicians have undertaken "to record *[d'enregistrai]* their experiences." With a broader base of empirical knowledge than the testimony of just three men, Montaigne remarks, we might "see some light" (763 [595).

Despite his rhetorical jibes at physicians, I submit that Montaigne has thus offered an agenda for advancing their profession. His remarks about the need for experiments to be conducted under a perfectly controlled "fortune" amount to a description of what we know of as the controlled experiment. And as for the need to broaden the scope of the "experience" on which medical findings are based, this is precisely what the essayist purports to do in most of the last two-thirds of his final chapter, titled "Of Experience": to offer a sample record of his eating, sleeping, and sexual habits, along with his experiences with the kidney-stone.[18] He introduces the account by describing the *Essays* as a whole ("this fricassee") as "a record [*registre*] of the essays of my life" (III.13, 1056 [826]; besides the verbal connection to the "record" for which he had called in II.37, note that the term "*essai*" is also translatable as "trial" of "experiment"). While deprecating the book as "exemplary" for one's "internal [spiritual?] health" only if one takes its instruction "in reverse," he adds that nobody offers "more useful experience" with regard to "bodily health," since he presents it "pure, not at all corrupted or altered by art" or opinion. And he emphasizes that experience rather than reason is the true foundation of medicine (1056 [826]).

The opening pages of III.13 articulate the epistemological foundations of Montaigne's pursuit of knowledge derived from experience. He begins by defending that pursuit, despite its supposedly being a "weaker and less dignified" means than reason (whatever that might mean), on the ground that "truth is so great a thing that we must not disdain any medium that will lead us to it"—hardly the position of a dogmatic skeptic. He adds that experience, like reason, "has so many shapes that we know not which to lay hold of . . . The inference that we try to draw from the resemblance of events is uncertain, because they are always dissimilar; there is no quality so universal . . . as diversity and variety" (1041 [815]). Similarly, he adds a couple of pages later, "[n]ever did two men judge alike about the same thing, and it is impossible to find two opinions exactly alike, not only in different men, but in the same man at different times." Human beings are unaware, Montaigne remarks, of "the natural infirmity of their mind," such that "it does nothing but ferret and quest," but becomes "crossed by so many difficulties and obstacles, and diverted by so many new quests, that it strays from the road, bewildered." Nonetheless, as he had done in the "Apology," Montaigne derives a lesson of hope from this apparently hopeless situation in the following eloquent passage:

> It is only personal weakness that makes us content with what others or we have found out in this hunt for knowledge. An abler man will not rest content with it. There is always room for a successor, yes, and for ourselves, and a road in another direction. There is no end to our researches; our end is in the other world. It is a sign of contraction of the mind when it is content, or of weariness. A spirited mind never stops within itself; it is always aspiring and going beyond its strength; it has impulses beyond its powers of achievement. If it does not advance and press forward and stand at bay and clash, it is only half alive. . . . [I]ts food is wonder, the chase, ambiguity (1045, [817–18]).

Part of the reason for such hope in the face of nature's infinite variety, as we learn shortly thereafter, is that even though "no event and no shape is entirely like another, so none is entirely different from another. . . . All things hold together by some similarity; every example is lame, and the comparison that is drawn from experience is always faulty and imperfect; however, we fasten together our comparisons by some corner" (1047 [819]). In these remarks Montaigne espouses a constructivist account of human knowledge, anticipating the epistemology of Locke and Kant: even though we are never able to arrive at a finally valid picture of the universe, our faculty of arranging the information we receive from the senses into categories of our own devising enables us to advance in comprehending nature's mode of operation.[19] As Montaigne remarked in the "Apology," the very fact that "things do not lodge in us in their own form and essence," as is shown by the variability of our perceptions depending on our own condition, means that "external objects surrender to our mercy; they dwell in us as we please" (II.12, 545 [422]).

SKEPTICISM AND POLITICS

Montaigne's accounts of the prospect of scientific advance in II.12 and III.13 are each intermixed with treatments of political and ethical issues in a manner that elaborates the relation between his teachings about science and politics. In the earlier chapter, the same skeptical attitude that dispels us of the illusion that our perceptions disclose the true structure of nature is also said to dictate an attitude of moderation with respect to moral and religious judgments. Noting "the confusion that our judgment gives to our own selves, and the uncertainty that each man feels within himself," Montaigne urges that we "become wise at our own expense" rather than suppose, amid our ever-changing beliefs, that "the present and the latest one is the certain and infallible one." Whatever is "preach[ed] to us," we need to recognize that "it is man that gives and man that receives," rather than believe that any teaching comes directly "from heaven." Hence the appreciation of "our faulty condition" and the frequency with which our understanding "receive[s] false things" should at least "make us behave more moderately and restrainedly in our changes." Montaigne reports that having "sp[ied] on [him]self more closely" than others do and noted the "vanity and weakness" and "instability" of his judgment, he has engendered "a certain constancy of opinions" in himself, scarcely altering his "original and natural ones" (II.12, 546–48, 553 [423–5, 428]). He contrasts this moderation born of self-knowledge with the tendency of preachers, lawyers, and judges unconsciously to allow their passions to guide their judgment and belief, and suggests that "the ardor that is born of spite and obstinacy against the pressure and violence of authority, and of danger, or the concern for reputation, has . . . sent some men all the way to the stake to maintain an opinion for which, among their friends and at liberty, they would not have been willing to burn their fingertip" (547–550 [424–6]).[20]

In III.13, Montaigne infers from the infinite variability of circumstances that it is imprudent to attempt "by a multiplicity of laws to bridle the authority of judges," since "there is as much freedom and latititude in the interpretation of laws as in their creation," and the surfeit of French laws still leaves judges with "a powerful and licentious freedom" that is unprecedented (1042 [815]). Given the lack of relation between human actions, which are in "perpetual mutation," and "fixed and immutable laws," it would be preferable to adopt those laws that are "rarest, simplest, and most general," like those of nature. Curiously, despite his denunciation of the excessive license judges already enjoy, Montaigne asks what danger there would be in allowing disputes to be settled by "our wisest men . . . according to the circumstances and at sight, without being bound to precedents, past or future" (1042–3 [815–16]). But part of the ground for this suggestion is supplied by his denunciation of the manner in which excessive regard for the "forms" of justice causes innumerable innocent men to be punished—as when judges in the essayist's "own time" deliberated about whether certain individuals who had been condemned to death for murder should still suffer that punishment, even though in the meantime other prisoners had "confess[ed] outright" to the deed and shed "a decisive light on the whole business" (1047–8 [819][21]).Thus those "wisest men" are presumably not the established judges.

The practical bearing of the foregoing remarks is suggested by Montaigne's account of his own conduct and attitude in the sequel. The essayist professes to imitate Alcibiades in being unwilling to "turn himself over . . . to a man who can dispose of my head, where my honor and my life depend on the skill and diligence of my attorney more than on my innocence." After describing himself as "sick for freedom," Montaigne proceeds to express his determination, in contrast to those he cited in II.12 as prepared to accept martyrdom out of passion in defense of opinions for which they might not otherwise be willing to sacrifice a fingertip, to *abandon* the laws that he "serve[s]" if they "threatened even my fingertip" (1049 [821]; emphasis added). Hence Montaigne appeals beyond the laws of France, so "confused and inconsistent that they are some excuse for both disobedience and faulty interpretation," and which are "often made by fools, . . . who in their hatred of equality, are wanting in equity," to the laws of "nature." These are not the "rules of nature" described by philosophers, who "falsify them and show us the face of nature painted in too high a color, and too sophisticated," but the "general law of the world," which Montaigne will "know . . . well enough when I feel it" (1050 [821–2]).

Montaigne's remarks about the true laws of nature recall his reference in the preface to the "nations which are said to live in the sweet freedom of nature's first laws," among whom he would have been able to portray himself "entire and wholly naked"—whereas "respect for the public" has compelled him to be less straightforward (2). The nations in question are apparently exemplified by the cannibals portrayed in I.31, who indeed are said to go about naked and whose "whole ethical science contains only these two articles: resoluteness in war and affection for their wives" (206 [154]). Montaigne contrasts as well the absence of political and economic hierarchy among the cannibals with the artificial inequalities that exist among the French ([212–13 [159])—reminding us of his criticism in III.13 of those whose hatred of equality generates inequity. But for an elaboration of what it would mean to follow the laws of nature, we are ultimately brought back anew to the "Apology," where Montaigne initially mocks those who profess to describe "natural laws," remarking that "the only likely sign" of such laws would be "universality of approval," an unlikely thing in view of the infinite variety of customs and positive laws to be found in the world. While granting that there may be natural laws, "as may be seen in other creatures," Montaigne remarks that they are "lost" among us because "human reason . . . confus[es] the face of things with its vanity and inconsistency" (563–4 [437–8]). Yet only a couple of pages later, Montaigne proffers a description of a way of life that he judges to accord with *both* reason and nature, that of the philosophers (567–69 [439–440]).

Montaigne's account of the philosophers' way of life is singular in that it omits any reference to the pursuit of wisdom. Rather, he emphasizes the "excessive license" that "the soundest philosophy allows": the free indulgence in all sorts of sensual pleasure, in violation of customary standards of "decency," limited only by "moderation and the preservation of others' liberty." In unmasking "the freedom of philosophical opinions concerning vice and virtue" (567 [439–40]), Montaigne challenges the practice he attributes to the philosophers in III.12 and elsewhere of offering an excessively "sophisticated" portrait of nature for the public at large, covering up their hedonism and pursuit of their own good out of concern to offend, or corrupt, the multitude (who could not be expected to limit such indulgence by considerations of moderation as the philosophers did, lacking their wisdom).[22] In sum, in Book III Montaigne attributes to the classical philosophers as a whole the "bent stick" precept that Aristotle offers as his provisional moral advice in the *Nicomachean Ethics:* to steer people toward the extreme that is opposed to the one to which they naturally incline, the pursuit of pleasure, in the hopes of thereby approximating the mean (III.10, 983 [769]; III.9, 968–9 [757]; *Ethics* 1109b5–7). In opposition to that practice, Montaigne—in the spirit of Machiavelli—condemns the practice of setting goals that are beyond man's capacity to achieve, or of proposing "patterns of life which neither the proposer nor his hearers" have any hope or desire to follow (III.9, 967–9 [756–8]).[23] By contrast, the "law of nature" that he admonishes us to follow is one that is accessible through feeling (III.13, 1050 [821])—that is, it would appear, our animal sensitivity to physical pleasure and pain.[24]

We are now better equipped to address the relation among Montaigne's skepticism, his natural philosophy, and his political teaching. By comparison with the Pyrrhonian position that he mocks for its disregard of common sense, Montaigne's is an attenuated skepticism. While rejecting the assumption that information derived from our sense perceptions and accumulated experience gives us a true portrait of reality—given the variability of our perceptions and the gap between our finite experience and nature's infinite variety—Montaigne both emphasizes the necessity of grounding our everyday lives on our perceptions and intimates the possibility of constructing a science of medicine on the basis of systematic observation of nature. While that science may never provide us with a final, comprehensive picture of the universe, it is inherently progressive and should enable us to make human life progressively longer, healthier, and more comfortable.

Although Montaigne's emphasis on the analysis of sense data as the ground of a progressive natural science aimed at uncovering and hence manipulating nature's observation harmonizes with his hedonistic moral teaching, it cannot be said that the latter is the result of the former. To the contrary: Montaigne represents the human situation as allowing us to choose among a variety of more or less equally plausible accounts of nature, none of them susceptible of definitive proof: amid his account of the ancient philosophers' skepticism he cites Timaeus (from Plato's eponymous dialogue) as saying of his knowledge of the gods, the world, and men that "it is enough if his reasons are as probable as another man's," since "the exact reasons" are unavailable to him or to any other mortal (II.12, 487 [375]. More generally,

Montaigne says of such wise men as Epicurus, Plato, and Pythagoras that they propounded their metaphysical doctrines not as final truths, but as attempts to bring "some *image* of light" into "the obscurity and ignorance of this world" (II.12, 491–2 [379] [emphasis added]): whatever light we enjoy is a self-generated "image," not nature's own illumination (compare the account of the cave in Plato's *Republic*). Under these circumstances, faced with "our miserable human condition," Montaigne asks at the outset of I.14 (in language echoed in remarks quoted above from II.12), if external objects "give themselves up to our mercy, why shall we not dispose of them and arrange them to our advantage?" (I.14, 50 [39]).

Montaigne's adoption of a deterministic and materialistic account of nature appears to be more the consequence than the cause of his libertarian political teaching. Since previous philosophers' "sophisticated" portrayals of nature backfired, by providing an unintended ground for theocracy, Montaigne judges it preferable, in his time, to openly portray nature in purely material form, and as pointing us to no goal beyond sensual pleasure. If his predecessors purposely obscured the gap between nature's infinity and man's purposes, it may nonetheless be suggested that Montaigne himself obscures for polemical purposes something no less important. Nature in his portrayal seems to offer no ground for his own choice of a philosophic (and partly political) life.[25]

ENDNOTES

1. Part I of this essay borrows from my paper "Skepticism, Dogmatism, and the Future of Liberalism," presented at the 2008 meeting of the International Society for the Study of European Ideas.

2. See, e.g., Richard Popkin, *The History of Scepticism from Erasmus to Descartes* (rev. ed., New York: Harper Torchbooks, 1964), ch. 3. For dissenting views, see Elaine Limbrick, "Was Montaigne Really a Pyrrhonian?" *Bibliothèque d'humanisme et Renaissance*, 39.1 (1977), 67–80, and Alan Levine, *Sensual Philosophy: Toleration, Skepticism, and Montaigne's Politics of the Self* (Lanham, MD: Lexington Books, 2001), 37–38.

3. Parenthetic citations of the *Essais* refer to book, chapter, and page number in the Pléiade edition of Montaigne's *Oeuvres completes,* ed. Thibaudet/Rat (Gallimard, 1962), with the fourth, bracketed number identifying the page in Donald Frame's translation (Stanford University Press, 1957). For an extended analysis of the "Apology for Raymond Sebond" see my *The Political Philosophy of Montaigne* (Ithaca, NY: Cornell University Press, 1990), chs. 2–4.

4. On the nonphilosophic multitude as "children" from the perspective of the philosopher, cf. Plato, *Gorgias,* 521e.

5. The inapplicability of the account to Socrates is suggested by the references to the philosopher's having nothing to do with "dinners and celebrations" (compare Plato's *Symposium*); to his being unaware of his own ignorance; and to his neglecting "the things nearby" in pursuit of knowledge of the things "above the heavens" and "underneath the earth," precisely the "old accusations" that the Platonic Socrates (unlike the Aristophanic one) denies apply to him (*Theaetetus*, transl. Joe Sachs [Newburyport, MA: Focus Publishing, 2004], 173d–e; *Apology of Socrates,* 18b, 19b–c). See also Ronna Burger, *Aristotle's Dialogue with Socrates: On the "Nicomachean Ethics*" (Chicago: University of Chicago Press, 2008), 110.

6. *Nicomachean Ethics* 1139b35ff., 1141b19–20, 1141b2ff.

7. See Burger, *Aristotle's Dialogue with Socrates,* ch. 4. In citing the same passage from the *Ethics,* Montaigne is careful not to treat the opinion that wisdom and prudence are separable as Aristotle's own view, but observes only that he attributes it to "some" *(aucuns)* (I.25, 135 [99]).

8. Although in his initial presentation of the Pyrrhonian way of life as a reasonable one that accommodates "natural inclinations" and "laws and customs" (following Sextus Empiricus's description in *Hypotoposes* I.12), Montaigne observes that it is hard to reconcile with this doctrine Pyrrho's reportedly "wild and unsociable" behavior (II.12, 485 [374], in II.29 he treats such reports as accurate and describes Pyrrho as having constructed an "amusing science" out of "ignorance," hardly evincing thereby any great approbation of Pyrrhonism (683 [533]).

9. This aspiration contradicts the concern that Montaigne professes near the beginning of the "Apology for Raymond Sebond" that the Reformation, if unchecked, "would easily degenerate into an execrable atheism" once "the vulgar" were "given the temerity to despite and judge the opinions that they had held in extreme reverence" (II.12, 416 [318]). In fact, in that chapter, while purporting to defend Sebond's *Natural Theology* against its rationalistic critics, Montaigne actually upholds the cause of philosophy against that of revelation, as is shown in *The Political Philosophy of Montaigne,* chs. 2–4, and the studies by Zbigniew Gierczynski and Arthur Armaingaud cited therein. See also my essay "Montaigne, Tocqueville, and the Politics of Skepticism," *Perspectives on Political Science,* 31.4 (Fall, 2002), 204–212.

10. See Nietzsche's preface to *Beyond Good and Evil.*

11. Note that Montaigne's source for the figure of 288 sects that supposedly divided over the issue of the supreme good (II.12, 561 [435]) is the Roman philologist Varro (as reported by St. Augustine in *The City of* God XIX.11); see also Cicero, *Academica*, II.42. In other words, the seeming variety of philosophic opinions on the subject is by no means a modern discovery.

12. See Hobbes, *Leviathan*, ch. 11, first paragraph; Locke, *Essay Concerning Human Understanding,* ed. Peter Nidditch (Oxford: Clarendon Press, 1975), II.21.55, pp. 269–70.

13. See *The Political Philosophy of Montaigne,* ch. 1.

14. On the distinction between Pyrrhonian and zetetic or Socratic skepticism, see Stewart Umphrey, *Zetetic Skepticis*m (Wakefield, NH: Longwood Academic, 1990), 297–98. See also Leo Strauss, *On Tyranny* (Glencoe, IL: Free Press, 1963), 109–10.

Montaigne's renunciation of the Pyrrhonian goal of ataraxy is acknowledged by Frédéric Brahami in *Le Scepticisme de Montaigne* (Paris: Presses Universitaires de France, 1997), 61. Brahami nonetheless persists in regarding Montaigne as a Pyrrhonian.

15. See *The Political Philosophy of Montaigne,* 115–33.

16. See Pierre Villey, *Montaigne et François Bacon* (Geneva: Slatkine Reprints, 1973), and Etienne Gilson's edition of the *Discours de la méthode* (Paris: J. Vrin, 1947).

17 The French word *experience* translates as both "experience" and "experiment."

18. This record is buttressed, at an apparently more reliable or authentic level, in the author's *Travel Journal,* in which he recorded the apparent effects of visiting different baths on the kidney-stone. It is not known whether Montaigne intended for this journal ever to be published.

19. This presupposes a regular or deterministic account of nature, as suggested at II.12, 445 [343] and elsewhere: see *The Political Philosophy of Montaigne,* 284–85.

20. The theme last mentioned is also developed in I.14 (implicitly contrasting the readiness of the Portuguese Jews to sacrifice their children's lives as well as their own rather than allow them to be forcibly converted with Montaigne's own indifference to nonbodily pains) 52–5, 61 [35–6, 42] , I.30 (on the embrace in "every religion" of opinions that spur us to "massacre and homicide") (199 [149]) ; I.33, and II.2, 328–30 [249–51] (on the dangers of spiritual "drunkenness"), among other passages. See *The Political Philosophy of Montaigne,* 204–212 and ch. 8.

21. Cf. Hobbes's similar lament at the "barbarous" reliance on irrational precedent, *Leviathan,* ed. E.M. Curley (Indianapolis: Hackett, 1994), ch. 11, par. 24, 61; also Montesquieu, *The Spirit of the Laws,* 29.1 (on the danger of an excessive regard for the forms of legality).

22. More precisely, it should be noted that the philosophers he depicted as outright hedonists are different from the ones who set forth an elevated understanding of nature for the benefit of the multitude: the former—a motley group belonging to such sects as the Stoics, Cynics, and Peripatetics—are said to have mocked as "stupidity" "what we call decency—not to dare to do openly what it is decent for us to do in private" (such as farting and masturbating) (II.12, 567–68 [440–41]). Unlike the practitioners of an ostensibly edifying rhetoric, the overt hedonists were not *political* philosophers. The novelty of Montaigne's position lies in his combining a political project with a profession of "frankness" about nature, prefiguring the Enlightenment.

23. Lest one exaggerate the opposition between Montaigne and Aristotle on this point, we should note that his criticism of those who disparage pleasure only to turn the multitude away from its pursuit is anticipated by Aristotle, who late in the *Ethics* (X.1), following the elevation of intellectual virtue over moral habituation, implicitly renounces the bent-stick principle with respect to pleasure.

24. Cf. I.14, 55 [37] regarding the "general habit of nature which can be observed in all that lives under heaven that we shall tremble under pain"; the instinctive fear of pain is the means by which nature impels all animals towards their preservation (I, xiv, 55 [37]; III, xii, 1032 [807–08]) with John Locke, *Two Treatises of Government,*

II, secs. 94, 209, and 225, and Robert Goldwin, "John Locke," in Leo Strauss and Joseph Cropsey, eds., *History of Political Philosophy* (3rd ed., Chicago: University of Chicago Press, 1987), 506 on the role of popular feeling in Locke's political teaching. At II.8, 365 [279] Montaigne identifies the instinct of self-preservation as the most likely candidate for the status of a natural law, assuming that there is any such law.

25. Cf. Socrates's disappointment at Anaxagoras's failure to account for the role of mind in ordering the universe: Plato, *Phaedo*, 97b–99c.

Chapter 8

Parmenidean Intuitions in Descartes's Theory of the Heart's Motion

Dwayne Raymond

YEAST IN FOLDS OF THE HEART

In reply to a damning observation that the excised heart of an eel continues to pulse, Descartes suggests that there is "in the folds of the heart some humor resembling a ferment (or yeast)?"[1] What drives him to make this claim? Descartes reports that his interlocutor, the physician Plempius, suggested that the view is "quite far-fetched, and that [Descartes] was using it as a last resort, as if [he] were hard pressed, and the point provided the only way of escape."[2] Is Plempius right? Is the yeast conjecture little more than a desperate attempt to salvage a rationalist theory in the face of an empirical counterexample? That is, is Descartes some sort of village rationalist[3] who "did not control his theory by observation or experiment?"[4] Or, is Descartes a kind of empiricist[5] who "seems positively to glory in the face of evidence of the senses, in what can be seen and touched?"[6] Indeed, in reply to Plempius, Descartes holds that, while his view does not require the yeast conjecture, observations and theory both support it. He writes that:

> it is necessary to admit that some blood that is rarified remained in [the heart] from one diastole to the next, and that rarefaction is increased by the fresh blood that flows in and mixes with it; and that the nature and properties of fermentation are clearly relevant to this fact.[7]

The protracted debate over the heart's motion is a rich source of material documenting the Cartesian scientific practice of drawing upon *both* observation and theory. G. Groham has correctly noted that neither the village rationalist readings, nor the *purely* empirical readings adequately accommodate both dimensions.[8] His third interpretation brings these elements together. Shifting focus from method to metaphysics, Gorham maintains that Descartes method is shaped by his metaphysics.[9] While this approach provides the most robust explanation (accommodating both dimensions), it directly contravenes Descartes's repeated claims that his entire philosophy is the product of a method. Method is supposed to yield his particular metaphysics.

In what follows, I'll explain the yeast conjecture and draw upon an ancient intuition to argue that method drives Descartes's metaphysics. I will not characterize Descartes's method. My task is show that an ancient intuition, one that informs logic and aspects of ancient method, is explicitly at work in his discussions about the method. This intuition informs the means by which he arrives at his metaphysics which then contributes to his position.

THE CENTRAL ISSUE: WHAT CAUSES THE HARDENING OF THE HEART?

The Harvey-Descartes debate over the motion of the heart turns on divergent explanations of its hardening. Whereas Descartes maintains that the heart hardens while *expanding,* the English physician William Harvey maintains that it hardens while *contracting.*

In *de Motu Crodis*[10] Harvey held that the heart's motion results from a contractile element located in the heart. A "pulsific faculty" originates the initial movement in the heart's right auricle. This contractile movement is the first in a chain of four separate and distinct movements. The four movements occur in a four-chambered heart. Blood flows continuously into the right auricle by way of the *vena cava.* Upon reaching capacity, the auricle contracts, driving the blood into the right ventricle. The ventricle then contracts, driving the blood into the left auricle by way of the "arterylike vein" or pulmonary artery. The left auricle then contracts, driving the blood into the left ventricle. These four distinct but perfectly synchronous movements appear as two movements with two phases. In the systolic phase, the blood is expulsed. The heart whitens, becoming shorter and harder. It strikes the wall of the chest. In diastole the heart fills with blood. It reddens, becoming longer and softer.

Descartes, on the other hand, did not postulate an inexplicable pulsific faculty. His view requires only one type of motion, change in extension. In his *Descriptions of the Human Body* he holds that nutrified venous blood enters the *vena cava* from the liver. One (large?) drop at a time, it fills the heart's right chamber. At the same time, blood enters the heart's left chamber from the "pulmonary vein." There are only two chambers. Encountering a heat in the heart, a certain "fire without light," the blood is instantaneously rarified. During this process, venous valves are forced shut and arterial valves are forced open. The blood in the right cavity is driven into the pulmonary artery. The blood in the left chamber is driven into the arteries, simultaneously expanding them. The heart becomes longer and softer, once it has expelled the blood. This two-chambered heart hardens along with the blood's expansion.

To summarize both hold that blood is being expelled when the heart hardens. Harvey correctly maintains that its hardening is due to contraction. His view relies upon an inexplicitly pulsific faculty. Descartes held the erroneous view that its hardening is due to the blood's expansion from a process of instantaneous rarification. His view only employs the one source of motion: change in extension.

THE PULSING OF AN EXCISED HEART: TWO ROBUST RECONSTRUCTIONS

The first robust reconstruction pits Harvey's vitalist conception against Descartes's mechanist physics. Harvey's relied upon an inexplicable occult quality to explain the source of the heart's motion. Since Descartes's metaphysical physics will not admit occult qualities, Descartes is forced to reject Harvey's view as an "unknown or strange faculty." [11]

In this vein Descartes's yeast conjecture could be understood in terms of his metaphysical physics. But instead of offering such an account, M. Grene is simply astonished by Descartes's suggestion that a yeastlike ferment may lie in the folds of the heart.

She classifies the conjecture as one of five separate replies, each of which is supposed to independently salvage Descartes's view against the damning counterexample. In Grene's words, the excised heart continues to beat because:

> (1) There is always blood; (2) for the purposes of distillation, the less, the better; (3) the heart is used to dilating, and so goes on doing it **(which seems to contradict the first point)**; (4) perhaps there is "in the folds of the heart some humor resembling a ferment (or yeast), by mixing with which another humor, arriving, makes it [the heart] swell." [. . .] Finally (5), still answering Plemp's first objection, Descartes points out that his view contradicts and improves on, the common opinion of those who believe that the heart beat depends on some faculty of the soul.[12]

Seemingly unmotivated, these five independent replies are said to be jointly inconsistent. The third contradicts the first. Descartes does seem to be deeply confused; Plempius's characterization appears apt. The set of five seemingly inconsistent replies appear not to be driven by any method, but by desperation.

Whereas the foregoing first robust reconstruction dealt with the metaphysical physics, the second maintains that *both* mechanism and mind-body dualism are required to explain Descartes's rejection of Harvey's view.[13] Harvey's view seems "to require an unconscious mental operation, something entirely contrary to the principles of Descartes's dualistic ontology."[14] According to Gorham, Descartes only appeals to a nonphysical cause where conscious volition is immediately present. All other cases require a mechanistic explanation. "[T]his methodological rule follows directly from the dualistic metaphysics which confines the capacity for independent action to the realm of spirit."[15] Gorham is very emphatic about the dependency here. He also writes, "[t]o me it seems clear that Descartes's limited anti-vitalism and his resultant dispute with Harvey derives from his famous ontological dualism of mind and body."[16]

Recall Grene's list. Dualism is at issue in the *so-called* fifth reply. However, dualism *per se* does not explain the yeast conjecture. An appeal to Descartes's mechanistic framework is required. Unfortunately, Gorham has nothing to say on the matter. In a footnote, Gorham defers to Grene's analysis of the text. As we have seen, Grene's analysis is wanting.

In addition to this wanting analysis, Gorham's view that metaphysics drives Descartes's method, is immediately at odds with Descartes's own view. In what follows, I'll correct the wanting analysis and show that both Descartes's dualism and Law B result from his development of ancient intuition, showing that method drives Descartes's philosophy.

THE PULSING OF AN EXCISED HEART: REBALANCING THE QUANTITY OF MOTION

The yeast conjecture is not an independent reply within a set of five inconsistent replies. The fifth reply stands alone; it alone concerns the soul. The first four are unified by a single theme—transfer of motion. This is the basis for his metaphysical physics.

How does Descartes explain the continued pulsing of an excised heart? He rebalances the equation between the required motion and the available motion. Since the heart moves, the available motion cannot be less than the required motion. For this reason Descartes does two things: he decreases the amount required to move the heart and he increases the amount available *per* drop of blood.

If the heart pulses because of expanding blood, blood should be present. Grene's *so-called* first reply, in which it is observed that blood is present, indicates that a source of motion is present in the beating heart. This must be confirmed by observation. Descartes's *so-called* first reply asserts this. The equally *so-called* second reply continues in this vein. By observing that, for the purposes of distillation, less blood is better Descartes is increasing the amount of motion that is available per drop of blood. Grene's *so-called* third reply misrepresents Descartes point. She claims that Descartes held that the heart is "used to dilating and so goes on doing it." However, Descartes actually claims that, "as the heart expands and contracts a countless number of times from the first movement of its formation, the slightest force is enough to impel it to carry out its own repeated movement."[17] In his rebalancing of the equation, Descartes is reducing the required amount of motion to move the heart. Finally, the *so-called* fourth reply—*the yeast conjecture*—suggests an additional source of motion, should it be required. The presence of a yeast-like substance in the folds of the heart increases the amount of motion per drop of blood.

The yeast conjecture is part of a unified reply. The entire reply is obligated by Descartes's laws of motion. It is obligated by Law B, which balances the amount of motion received with the amount of motion available. According to Law B: When one body pushes another it cannot give the other any motion unless it loses as much of its own motion at the same time; nor can it take away any of the other's motion unless its own increases by as much.[18] Law B informs Descartes's overall reply and the yeast conjecture in particular.

A discussion of motion is required when the issue pertains to extension and changes in extension: the physical realm. A discussion of mind is required when an opponent suggests an explanation involving nonphysical elements, such as the soul. To this end, dualism informs, at least, the *so-called* fifth reply. If Gorham's analysis of Harvey is correct, it may also inform Descartes's rejection of Harvey's pulsific faculty. As we will see, Descartes's dualism and the laws of motion result from his unique development of an ancient intuition.

AN ANCIENT INTUITION

Even today the ancient intuition lingers in familiar figures of speech. We talk about getting to heart of the matter. We talk about the core issue. We say things like, "it all boils down to . . ." Each suggests a distinction between surface appearances and a deeper underlying reality. We get at this reality by stripping away the surface appearances, unveiling the heart of the matter. Such cases are best described as instances in which the truth is discerned. To discern what is true is to distinguish it from what is false; it is to clarify things by removing obscurities, leaving behind that which is clear and distinct.

The intuition, call it inseparability, is suggestive of a simple, yet powerful means of separating the deeper reality from surface appearances. To distinguish between the attributes which must belong to a subject from those that have no requirement to belong, simply remove the elements. Those that unnecessarily belong can be removed with no consequence to the underlying subject; they are separable. Those that must be present cannot be removed without destroying the underlying subject; they are inseparable. Inseparability concerns a nonreciprocal dependency relationship.[19] This basic relationship is crucial to working out both the essence of a substance, as well as its position within an internetwork—the *so-called* order of nature.

In *Topics VI,* 4, for example, Aristotle relies upon a co-demolition test (a test informed by inseparability) to establish priority. The genus is *prior* to the species because the species cannot continue to exist without it. Aristotle writes, "[f]or the genus and differentia destroy the species, so that these are prior to the species." Priority relations naturally developed into an interconnected network of dependencies: the *so-called* order of nature. Tuominen nicely characterizes the kind of relational epistemologies characteristic of Platonic-Aristotelian frameworks: "reality is taken to have an intelligible structure with discrete elements called forms. The forms have necessary connections with each other and the structure involves relations of priority. Therefore, reality itself has an intrinsic order. Knowledge in the proper sense is taken to be knowledge about the structure of reality."[20]

There are at least two ways that the ancients used inseparability to establish an interconnected network. Both are at work in Euclid. The first, as noted above, involves the unidirectional relation of a nonreciprocal dependency. It was illustrated above. The second involves a bi-directional or reciprocal dependency. [Descartes's laws of motion are structured by reciprocal dependency relations.] To test for this one first shows that A is nonreciprocally related to B and then one shows that B is nonreciprocally related to A. Combined the two tests establish a reciprocal relationship. This is exactly what Aristotle does in *Categories* 7, where he discusses co-relatives. [21] Consider his example of half and double. The half and the double must coexist. Why? The destruction of half entails the destruction of double. Why? To be half is to be half of something, namely it is to be half of what is double of it. To be double of something is to be double of what is half of it. The loss of one brings about the loss of the other. Something is not a double without being double of something. In this way, this relationship is reciprocal; the attributes half and double are each said to be co-relatives.

The centrality of this relationship in Euclid's Book I is often overlooked. Indeed, inseparability informs *both* Euclid's basic entities as well as two distinct relations between these entities. He defines a point as "that which has no parts."[22] That is, there is nothing separable from a point. It is homogeneous and whole. [As an aside, Szabo has links this notion of a point with Parmenides requirement.[23]] Unlike a point, a line has an attribute. "A line is breadthless length."[24] And "A surface is that which has length and breadth only."[25] There is a nonreciprocal dependency between point, line, and surface. A point can exist without a line, but line is dependant upon a point. A line can exist without a surface, but a surface cannot exist without a line. These nonreciprocal dependencies establish a hierarchy among these basic geometrical elements. The removal of, for example, breadth yields a line and the further removal of breadthless length yields a point. When one reaches the point, there is nothing further to remove. This is the starting point, the base upon which all else depends. (In Descartes, there are two such items: extended thing and thinking thing. As we'll see in section VI they are approached the same way, by removing what is apparent and retaining what is real. Descartes discerns the real.)

While the hierarchy establishes an ordering among the elements, it does *no work* in Euclid's proofs. His proofs rely upon the bi-directionality of co-relatives. In Euclid co-relatives are coexisting attributes or properties that always exist as a pair, if they are to exist at all. The emphasis is not the relation *per se;* it is upon the two entities which coexist. In Euclid, for example, there is a pairing between two intersecting lines and the angle at which they intersect. The intersecting lines and the angle at which they intersect are different aspects which must coexist. This is not a minor, nor is it an accidental feature of Euclid's writing. It is basic to his thinking about geometry. As David Reed has observed,[26] Definitions 13 and 14 jointly specify a co-relative relationship between figure and boundary:

> A boundary is that which is an extremity of any kind.[27]
> A figure is that which is contained by any boundary or boundaries.[28]

A boundary contains a figure and a figure is contained by a boundary. To have a boundary is to have a figure and to have a figure is to have a boundary. This kind of co-relative pairing structures each of Euclid's definitions of

Table 8.1

Boundary	Figure
Two points	Line (breadthless length)
A boundary line such that all straight lines falling upon it from one point among those lying within the figure are equal to one another	Circle

the figures. For example, every time the figure circle exists, it is contained by a boundary.

> "A circle is a plane figure contained by one line in such a way that all straight lines falling upon it from one point among those lying within the figure are equal to one another." [29]

While the above definitions for a point and a line are not related in this way, Euclid's third definition bridges this divide, stating the co-relative link between them: "The extremities of a line are points."[30] Since an extremity is a boundary, whenever two points exist, there is a breadthless length between them. In this way the early definitions of point, line, and surface are brought into line with Euclid's basic co-relative structure. But there is a problem here. While two points are co-relative with a line, the boundary is not specific to any one kind of line. The line could be straight or curved. This suggests a role for the construction postulates. They state the range of cases under consideration. Since the first postulate limits our attention to straight lines, two points are co-relative with a unique type of line—a straight line.

While it may seem odd to us that a line is a figure, or that an angle is a figure, it is worth keeping in mind that the figure boundary co-relative pairing provides for a nonhierarchical structure (in opposition to Aristotle's genus-species relationship[31]) among the elements. The pairings include pairings between highly abstract components such as two points and breadthless length. It plays a specific role in Euclid's demonstration. Euclid's demonstrations rely upon diagrams. Co-relative features contribute to his reasoning in at least two ways: first, they contribute to the universal character of his diagrammatic reasoning by limiting attention to the most generalized aspects of the diagrams; second, they render a substantial portion of his reasoning as bi-directional. Let's begin with the first. Each co-relative pairing restricts relevance to highly universalized features of a given diagram. Relevance is limited to that which is specified in the pairing. To see how this works, consider the proof for Proposition 1.1. Recall that co-relative pairings create contexts in which comparisons can be made, common notions enable the comparisons and the postulates specify a range of cases under consideration.

"Let AB be the given finite straight line. Thus it is required to construct an equilateral triangle on the straight line AB. With centre A and distance AB let the circle BCD be described; again with the centre B and the distance BA let the circle ACE be described; and from the point C, in which the circles cut one another, to the points A, B let straight lines CA, CB be joined. Now, since the point A is the centre of circle CDB, AC is equal to AB. Again, since B is the centre of circle CAE, BC is equal to BA. But CA was also proved

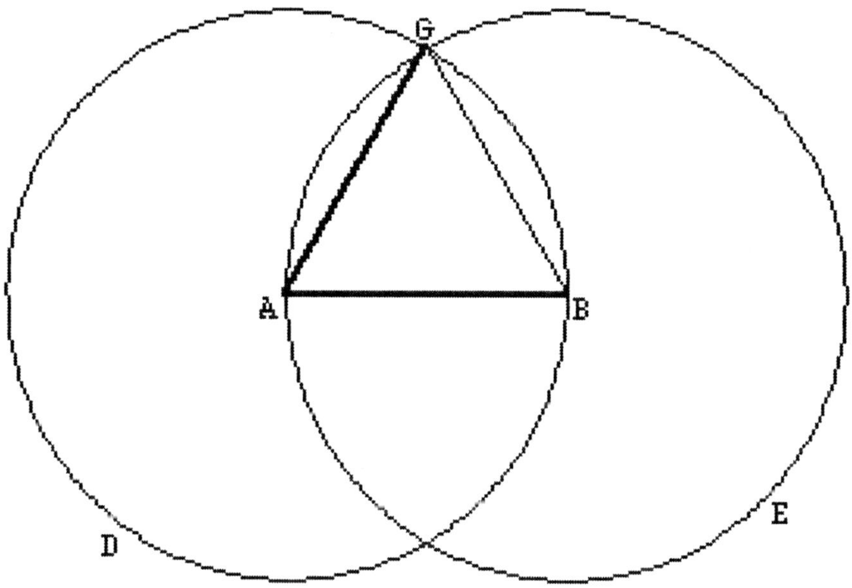

equal to AB; therefore each of the straight lines CA, CB is equal to AB. And things which are equal to the same thing are also equal to one another; therefore CA is also equal to CB. Therefore the three straight lines CA, AB, BC are equal to one another. Therefore the triangle, ABC is equilateral; and it has been constructed on the given finite straight line AB. (Being) what is required to do."[32]

The mathematically relevant features of the diagram are those which are specified by the following pairings.

The proof begins with a line whose points are AB. The objective is to construct an equilateral triangle. By constructing a circle with A at the centre and B on the circumference, the co-relative boundary condition for any given circle coexists. That is, in drawing a circle we also created a context in which lines connecting the center and any point on the circumference are equal to one another. The first circle is called CAE. By constructing a second circle, CDB, with the B at the centre and A on its circumference, a second context coexists. Since it is taken for granted that this construction is on the same plane, the two circles intersect. At the intersection we take point C. We now have two additional set of two points. Given two points a breadthless length coexists between them. The lines CA and CB are joined, completing the respective pairings.

At each stage in this construction, co-relative pairings dictate its construction. The members of various paring have various properties. Length coexists between two points and a context that equates certain lines which coexists with the figure circle. While the context establishes that line AB equals AC and that CB equals AB, the context cannot establish that AC equals CB. This equivalency is established by means of common notion 1. As Reed observes, the common notions do not establish contexts, "they relate measured things to each other."[33] The contexts enable measuring. And contexts emerge *via* co-relative pairings.

The scale of the particular diagram, the thickness of the lines, even the accuracy of the sketch—including the straightness of the lines, the shape of the circle, the presumed location of the circle's centre—all of these play no role in the demonstration. The highly abstract definitions and the co-relative pairings exclude both particular, nonessential features, and inexact renderings from having a role in the demonstration. The accuracy of the diagramed circle is not at issue. If CDB is a circle, it coexists with its boundary. That boundary contains the context which, in this proof, equated pairs of lines. The particular diagram aided the study by drawing attention to geometrical facts. The tight reasoning resulted because of the co-relative pairings and the contexts to which they gave rise.

Co-relativity provided for the necessity of the links. And the highly abstracted items paired contribute to the repeatability of the proof. Proposition 1.1 holds for the construction of an equilateral triangle on any length of line.

Moreover, given that each step involves a reciprocal relationship, one can begin with the finished equilateral triangle and work backward to a line of a certain length. At each vertex two lines of equal length meet. By placing the center of a circle at one vertex, each line falls upon the circle's circumference. By drawing a second circle in the same way at either remaining vertex, a different pair of lines begins at the centre and falls on the new circumference. Moreover, we notice that the two circles share a line and that pairs of lines meet exactly at one of the points in which the two circles intersect. Given that the two circles share a line, we can trace the entire procedure back to a line of a given length. If desired, we can further trace this line back to its two extremes or end points. The fact that the reciprocal links allow us to decompose (in analysis) the figure in this way and the then reverse the process and build a proof (in synthesis) does not cause a problem about priority. For as we have seen, the geometrical elements also have hierarchical relationship.

The fact that the basic geometrical elements are bound together by two networks (a network of nonreciprocal dependency relations establishing a hierarchal ordering, and a network of co-relative pairings, within an interconnected network of reciprocities) provides for the basis for explaining the link between analysis and synthesis. The description presented above is in keeping with Pappus's discussion of analysis. He describes it as, "the way from what is sought—as if it were admitted—through its concomitants in order to something admitted in synthesis."[34] Indeed, Hintikka and Remes argue that "akoluthein," which is usually translated as "consequences" (a nonreciprocal dependency relation), is given a better sense with "concomitants" (reciprocal dependency relation).[35]

We have seen that the geometrical definition of a point as "that which has no parts," is the product of an analysis which employs inseparability. The negative definition is characteristic of the approach. Elements are removed until no further removable elements remain. The interconnections between the elements are not always the same. Whereas Aristotle opts for a hierarchical structure like that of genus and species, Euclid differs. Euclid draws upon co-relatives to inform his reasoning and he relies upon the nonreciprocal dependency relations to establish a hierarchical structure between the geometrical elements.

In the next section, I'll claim that Descartes explicitly relies upon inseparability to arrive at his two basic substances, extended thing and thinking thing. Both are part of an interconnected network. Descartes uses quantity in terms of laws of motion to bridge the gap between two successive configurations of extension. The preservation of motion across an exchange was the key insight to Descartes puzzling remark that yeast may lie in the folds of the heart. The amount of motion prior to the exchange is equal to the amount of motion after the exchange. They are the Cartesian equivalent to Euclid reciprocal relation.

INSEPARABILITY WITHIN DESCARTES

Descartes relies upon inseparability to arrive at his two basic substances and to ground the quantitative relations known as his laws of motion.

Not only does Descartes's conception of reality contain an intrinsic order, but also the method by which he arrives at an understanding of this ordered reality explicitly employs inseparability. In *The Search for Truth* Descartes claims that:

> All truths are interconnected *(se invicem consequuntur)* and are mutually held together by a bond *(& mutuo inter se vinculo continentur)*. The whole secret is [a] to begin with the first and simplest truths, and then [b] to proceed gradually and as it were step by step to the most remote and most complex truths *(a primis et simplicissimis incipiamus, & deinde sensim & quasi per gradus usque ad remotissimas & maxime compositas progrediamur)*.[36]

Inseparability informs the means by which Descartes arrives at his basic substances and the means by which he identifies their interdependencies—they are held by necessary bonds. Inseparability is all about testing for such bonds. In his study of basic substances, Descartes relies upon the intuition. Consider his assertion in the second *Meditation:* "[a]t last I have discovered it—thought; this alone is inseparable from me."[37]

Inseparability is equally at work in the discovery of extension as a substance. In the second *Meditations* he writes:

> I put the wax by the fire, and look: the residual taste is eliminated, the smell goes away, the colour changes, the shape is lost, the size increases; it becomes liquid and hot; you can hardly touch it, and if you strike it, it no longer makes a sound. But does the same wax remain? It must be admitted that it does.[38]

All of these aspects are separable from the piece of wax. The wax remains without them. The second *Meditation* ends with the lesson that "bodies are not strictly perceived by the senses or the imagination but by the intellect alone."[39] While this lesson tells us how to make proper use of observations, it is not until the sixth *Meditation* that we have enough background to learn that the essence of body is extension. It alone is inseparable from an existing body, an extended thing.

The intuition informs part of the method by which Descartes arrives at dualism. It ultimately informs Law B. In *Le Monde* Descartes explicitly grounds the laws of motion in God's simplicity and immutability.[40] Law B requires that the quantity of motion be preserved through all changes. Descartes tells us that this law "follows manifestly from the fact that God is immutable and that, acting always in the same way, he always produces the same effect."[41] Inseparability informs God's attributes, such as immutability. In the third *Meditation* he equates simplicity with that which is inseparable—thus linking a method with the result.

The unity, *the simplicity or the inseparability of all the attributes of God* is one of the most important of the perfections which I understand him to have. And surely, the idea of the unity of all his perfections could not have been placed in me by any cause which did not also provide me with the ideas of the other perfections; *for no cause could have made me understand the interconnection and inseparability of the perfections without at the same time making me recognize what they were.*[42]

PARMENIDEAN INTUITIONS

There is a deep connection between Parmenides and Descartes. Both rely upon inseparability to pick out their basic ontological substance and both rely upon it to establish a necessary connection. The connection may be *via* geometry, as Arpad Szabo explicitly argues for the link between the definition of point as "that which has no parts" and Parmenides's homogenous whole.[43]

In the 5th century BCE, Parmenides employs inseparability to discover "[t]he unshakable heart of a well-rounded truth" in his famous poem, *On Truth.* [44] The metaphor is still with us and it speaks volumes. It suggests that there is an underlying reality, beneath the surface. Strip away the appearances until all that remains is the unshakable heart. As we saw above, Descartes also links simplicity with inseparability.

In Fragment 2 Parmenides uniquely develops inseparability into a method that generates a homogeneous whole.[45] In this fragment he discusses methods, presenting the only two possible paths of inquiry. He endorses the first path (the path of is) and rules out the second path (the path of is not) as being impossible. While he ultimately endorses a monism, which precludes an interconnected network of forms, his argument explicitly relies upon inseparability to establish an unbreakable chain.

He characterizes the first path of inquiry as the path of "'is,' and that it is not possible not to be." [46] According to Cordero, Parmenides is asserting the impossibility of separating is and being. While there is much debate over what Parmenides means by these terms, the crucial move in his argument involves establishing that they are inseparable; whatever the terms mean, they are inseparable. Given their inseparability, the second path is said to generate a contradiction. The second path of inquiry is the path of "'is not,' and that it is necessary not to be." [47] Since is and being are inseparable, the "is" in the phrase "is not" entails being, and the "not" denies it. Cordero nicely summarizes the contradiction: requesting that one to think of "that being that exists does not exist."[48] This is inherently contradictory. However we understand the meaning of "being" and "is," the fact remains: Inseparability plays a pivotal role in establishing that unshakable heart of his well-rounded truth. For Parmenides there is a single homogeneous whole. For Descartes there are two basic substances. For both philosophers the path to simplicity is *via* inseparability.

The yeast conjecture is not, as Plempius, suggested a "far-fetched, last resort." It is in keeping with Descartes's metaphysical physics. His conjecture is informed by Law B. Further, the distinction which Gorham relies upon—the distinction between extended thing and thinking thing—is ultimately grounded in a method, one which relies upon the inseparability intuition. Descartes's dualism of thinking thing and extended thing, as well as Law B are the product of his unique redevelopment of an ancient intuition, one which informs tests for nonreciprocal dependencies and reciprocal dependencies. The ancient intuition factors centrally into the geometrical methods analysis and synthesis. It is used to pick out the starting points of knowledge, a hierarchy among the elements and (in geometry) set of reciprocal relations among them. Descartes relies upon the intuition to establish his starting points and he develops a quantitative means (*via* Laws of Nature) to bridge changes in motion. Descartes's use of Law B to preserve the quantity of motion in a transfer of motion provides a quantitative equivalent to Euclid's use of reciprocal dependence between the co-relatives of figure and boundary. Both provide for a symmetrical relation between things they join. As we have seen, Law B provided the needed insight to explain Descartes's yeast conjecture.

ENDNOTES

1. Letter to Plempius, March 1638 cited in Grene, M., "The Heart and Blood: Descartes, Plemp and Harvey," in *"Essays on the Philosophy and Science of Descartes,"* ed. S. Voss (Oxford: O.U.P., 1993), 327.
2. Grene, "The Heart," 327.
3. Hall, T. "Descartes's Physiological Method: Position, Principle, Examples" *Journal of the History of Biology* 3(1970): 53–80; Sloan, P. R. "Descartes, the Skeptics, and the Rejection of Vitalism in Seventeenth-Century Physiology" *Studies in the History of and Philosophy of Science*, 8 (1977):1–28; Wear, A. "Harvey and the Ways of the Anatomists" *History of Science* 21 (1983): 223–49; Westfall, R. *The Construction of Modern Science: Mechanisms and Mechanics.* (Cambridge: C.U.P., 1971); Whitterridge, G. *William Harvey and the Circulation of the Blood.* (London, Elsevier, 1971).
4. Witterridge, *William Harvey,* 155.
5. Toellner, R. "The Controversy Between Descartes and Harvey Regarding the Nature of Cardiac Motions" in *"Science, Medicine and Society in the Renaissance: Essays to Honor Walter Pagel"* ed. A. Debus (New York: Science History Publications, 1972): 71–89; Gilson, E *Etudes de philosophie medievale* (Strasbourg, France: Université de Strasbourg, 1921), 192–205.
6. Grene, "The Heart."
7. Letter to Plempius, March 1638. in Descartes, R., *The Philosophical Writings of Descartes.* Volumes I, II, III. Trans. J. Cottingham, *et al.* (Cambridge: C.U.P., 1984), Vol. III, 96.
8. Gorham, Geoffrey "Mind-Body Dualism and the Harvey-Descartes Controversy" *Journal of the History of Ideas,* 50 (1994): 211–34.
9. Gorham, "Mind-Body," 231.
10. Harvey, W. *The Circulation of Blood and Other Writings*, ed. K. J. Franklin (London: Everyman's Library, 1963).
11. Descartes, *"Vol. I,"* 318.
12. Grene "The Heart," 327.
13. Gorham, "Mind-Body."
14. Gorham, "Mind-Body," 212.

15. Gorham, "Mind-Body," 231.
16. Gorham, "Mind-Body," 229.
17. Descartes, *"Vol. III,"* 80.
18. Descartes, *"Vol. I,"* 194.
19. Cleary J. *Aristotle on the Many Senses of Priority* (Carbondale, Ill: Southern Illinois U.P., 1988); Cleary J. *Aristotle and Mathematics: Aporetic Method in Cosmology and Metaphysics* (New York: E.J. Brill, 1995).
20. Tuominen, M. *Apprehension and Argument: Ancient Theories of Starting Points for Knowledge* (Dordrecht, The Netherlands: Springer, 2007), 6.
21. Aristotle presents a more robust account, allowing for different types of co-relatives in both *Categories* 7 and *Metaphysics* V 15.
22. Euclid *The Thirteen Books of the Elements vol I, II, III*, trans. Sir T. Heath (New York: Dover, 1956). vol. I, 153.
23. Szabo, A *The Beginnings of Greek Mathematics* (Reidel, 1978).
24. Euclid, *Elements vol. I,* 153.
25. Euclid, *Elements vol. I,* 153.
26. Reed, D. *Figures of Thought* (New York: Routledge, 1995), 12. I am indebted to Reed for his emphasis on context and measurement. While I make several changes to his conception of co-relatives, I takeover the bulk of his conceptual framework concerning measurement.
27. Euclid, *Elements vol. I*, 153.
28. Euclid, *Elements vol. I,* 153.
29. Euclid, *Elements vol. I,* 153.
30. Euclid, *Elements vol. I,* 153.
31. This may help explain why Aristotle's syllogistic is inadequate for Euclid's Geometry. Aristotle's logic employs nonreciprocal dependency relations, Euclid's reasoning makes use of reciprocal dependency relations.
32. Euclid, *Elements vol. II,* 241.
33. Reed, *Figures,* 19.
34. Hintikka, J. and Remes, U. *The Method of Analysis: Its Geometrical Origins and Its General Significance* (Boston: Reidel, 1974), 9.
35. Hintikka, *The Method,* 8–19
36. Miles, M. *Insight and Inference: Descartes's Founding Principle and Modern Philosophy* (Toronto: University of Toronto Press, 1999), 253.
37. Descartes, *"Vol. II,"* 19.
38. Descartes, *"Vol. II,"* 20.
39. Descartes, *"Vol. II,"* 22.
40. Descartes, *"Vol. I,"* 92–98.
41. Descartes, *"Vol. I,"* 96.
42. Descartes, *"Vol. II,"* 34.
43. Szabo, *Beginnings.*
44. Fragment 1, 28. All Parmenides translations by Cordero, 2004. Cordero, N. *By Being, It Is* (Parmenides Publishing, 2004); Curd, P. *The Legacy of Parmenides: Eleatic Monism and Later Presocratic Thought* (Parmenides Publishing, 2004).

45. There is much debate concerning the nature of Parmenides's monism. The details do not concern us here. For the present purposes, it is significant that he employed inseparability to generate *some* kind of monism.

46. Fragment 2.3.

47. Fragment 2.5.

48. Cordero, "*By Being,*" 77.

Chapter 9

Hobbes's Natural Condition and his Natural Science of the Mind in *Leviathan*

Paul Ulrich

Scholars who work on the writings of Thomas Hobbes do not often have occasion to comment on the "surprise" at the beginning of *Leviathan,* his most famous work on politics: Hobbes does not begin that book with a treatment of any topic directly related to politics; rather, he "commences his work with the genesis and function of sense perception" and then expands into a discussion of other elements of what we can call the inner life of human beings, such as memory, dreams, imagination, and thinking, and he does so using only terms from natural science, describing matter in motion.[1] Scholars explain this beginning point in appropriately Hobbesian terms, referring to the "polemic" Hobbes waged against the Aristotelians of his time and "the political agenda that underlies [the polemic]."[2] And they do so on very solid ground, since Hobbes himself explains that the Aristotelians are in charge in the universities in his time and are used by the Church to deceive ordinary people on the matter of what kinds of forces are at play here in the world. The Church makes use of Aristotelian arguments and terms to convincingly teach that spirits, angels, and other kinds of immaterial beings are in this world to such an extent that they should be feared and heeded more than earthly political powers (II, 8; XLVI, 13, and XLVI, 18).[3] Hobbes argues that if people properly understand the way humans perceive, and that it is possible to perceive only what is material, then they will be "more suited to civil obedience" and much more political peace will be achieved. In short, Hobbes's discussion of perception serves political purposes central to *Leviathan* as a whole.

As crucial as this role is that Hobbes's discussion of perception plays, it seems that it is not the only role it plays, especially considering that the topic of perception necessarily leads into a broader consideration of the inner life in general. While the explanation of the importance of perception to Hobbes just presented is in keeping with some of the political intentions of *Leviathan,* he may have had another intention as well. To see that intention, it is necessary to start not with chapter one, "Of Sense" but with what would more correctly be called the beginning of the book, the Introduction, the majority of which is on "reading" oneself and "reading" others through oneself.[4] That is, most of the Introduction is on studying the inner life of humans, their passions, opinions, designs, and so on, and in one way that study is more important for politics than, for example, the study of principles of justice, or law, or war. At the end of the Introduction Hobbes says, "He that is to govern a whole nation must read in himself, not this or that particular man, but mankind." Hobbes apparently identifies only one thing in *Leviathan* that a ruler must study and so it would not be correct to say that, overall, he intends the book to be anything like a handbook for princes. On the other hand, the unique identification of this one topic that a ruler must study attracts the reader's special attention.

Having caught our attention with this clear statement of what a ruler should make the object of study, however, Hobbes is rather less than clear on whether it is possible to conduct such study successfully and even on exactly what that study is. For example, in the sentence just quoted above, Hobbes goes on to say that this kind of study is "harder than to learn any language or science," and had already said just before that sentence that even if "one man read another by his actions never so perfectly, it serves him only with his acquaintances, which are but few" (Introduction, 4).[5] But while Hobbes seems to be suggesting that it is all but impossible to succeed in reading mankind through reading oneself, he nevertheless also suggests that in some way he has succeeded in reading all of mankind. The Introduction to *Leviathan* concludes: "yet when I shall have set down my own reading orderly and perspicuously, the pains left another will be only to consider if he also not find the same in himself. For this kind of doctrine admitteth no other demonstration" (Introduction, 4). Hobbes in effect claims to know something about all of mankind since he claims to be able to show anyone who reads him that he is saying something true about him (the reader of Hobbes's book). But this is not the same as claiming to be able to show someone that he knows something that is true about everyone, since one can find out whether what Hobbes says is true only by looking within himself to see whether he can find what Hobbes says is there (though the reader can actually see it only there, in himself). To put it differently: Hobbes says that one who will govern a nation needs to study all of mankind; Hobbes seems to know something about all of mankind; but Hobbes can demonstrate what he knows only to one individual at a time, and he can show an individual only that what he says is true about that individual; he does not claim to be able to show anyone that what he says is true about all mankind, for it is not even possible to prove his point in any way other than by having each individual look inside himself to see whether what Hobbes says is true about him (the individual, not Hobbes).

All of this is quite baffling, for it is not at all clear what kind of study could fit the account Hobbes gives. What is more, none of this is obviously of immediate, direct relevance to anything else of importance in *Leviathan* and so it would be tempting to pass over it all.[6] Yet Hobbes must have had a reason for placing this discussion so prominently in the Introduction and for saying that the one and only study of one who will govern a nation must be the inner life of all men. In what follows I will try to make some sense of this by examining the most famous part of *Leviathan,* chapter thirteen, "Of the Natural Condition of Mankind, As Concerning Their Felicity and Misery." Turning to this chapter is not arbitrary. It is the only chapter in *Leviathan* the title of which tells us that it will explain something about all of mankind, and Hobbes said in the Introduction that a ruler must study all mankind. Also, while Hobbes does not say in so many words that a ruler must understand the natural condition of mankind *per se,* it would seem that a ruler's primary task, in Hobbes's view, is maintaining peace or security, and so a ruler would need to understand the natural condition of war in order to understand the way to peace (XIII, 13 and XIII, 14). It seems sensible to look for a connection between that and the study Hobbes does say a ruler must engage in.[7] And, finally, a sure test of whether a particular topic is in fact important is to see if that topic has a key place in what appears to be the core of Hobbes's political teaching. And so I now turn to chapter thirteen on the natural condition of mankind.

The first question to ask about Hobbes's natural condition is "What is it?", and the obvious answer to begin with is that the natural condition is a condition of war. It is reasonable, thus, to assume that the natural condition according to Hobbes is an actual, observable and observed state of conflict among human beings, of the kind that his contemporaries would have been familiar with in the English Civil War. That assumption is all the more reasonable given what Hobbes says in the last paragraph of the entire book, that his writing *Leviathan* was "occasioned by the disorders of the present time" (A Review and Conclusion, 17). Additionally, Hobbes's response to something like an imagined objection to his account of the natural condition seems to make clear that he has in mind the actual, known relations among humans: "It may peradventure be thought, there never was such a time nor condition of war as this; and I believe it was never generally so, over all the world. But there are many places where they live so now" (XIII, 11). Hobbes goes on to specify "the savage people in many places in *America*" and makes reference to "the manner of life which men that have formerly lived under a peaceful government use to degenerate into, in a civil war" (XIII, 11). And what are probably the most famous words in *Leviathan* would seem to confirm further that the natural condition is an actual, real life possibility, for it is a condition of war of "every man against every man" (XIII, 8) and one in which "the life of man" is "solitary, poor, nasty, brutish, and short" (XIII, 9).

Yet there are multiple reasons to think that the natural condition Hobbes describes is not necessarily or even primarily an account of humans in actual conflict with one another, nor necessarily or primarily even an account of human interaction. His initial description of war is merely "the *time* men live without a common power to keep them all in awe," and he spells out that "WAR consisteth not in battle only, or the act of fighting," and, more strongly, "the *nature* of war consisteth not in actual fighting but in the known disposition thereto during all the time there is no assurance to the contrary" (XIII, 8; italics for emphasis not in original). War of course can be a time of actual fighting, but it need not be. It is first of all the mere absence of a common power, and that is because the "nature" of war is not hostile action but a hostile disposition, which is not always acted upon.

What, then, to make of those well-known words of Hobbes quoted above that do seem to say that the natural condition is one that can be observed among savages and in times of civil war? Those are apparently examples of times of war, but they are not necessary for Hobbes's account. As already mentioned, when Hobbes gives his examples of people in America and of formerly peaceful men who are in a civil war with one another, he is in effect responding to an imagined question or objection. He had already made clear that he had completed his account of the natural condition well before that question or objection, and that his account was not based on observations of humans living without government but rather was an "inference made from the passions" (XIII, 10), and not necessarily the passions of men living without government. As to the natural condition being a war of "every man against every man," Hobbes cannot mean that the natural condition of war exists only where literally every man is against every man. In any civil war one could name, individuals join sides, factions, parties, and so on, and those groups of individuals war against each other; an actual civil war is not a war of every individual against every individual. Hobbes's example of war found "in all times," of "kings and persons of sovereign authority" (XIII, 12), is, oddly, not an example of entire nations under those sovereigns in a natural condition of war with one another, for Hobbes says only that the sovereigns themselves are at war. But when explaining the causes of conflict found in the natural condition, Hobbes refers to men attacking others "with forces united" (XIII, 4) and to men taking offense at what is said or indicated regarding "their kindred, their friends, their nation" (XIII, 7). The natural condition of war can exist not only where united armies exist, but where there are families, friends, and nations, and so this too indicates that the natural condition of war need not be a war of literally every man against every man. What is more, when sovereigns are in a condition of war with one another "they uphold thereby the industry of their subjects, [and] there does not follow from it that misery that accompanies the liberty of particular men" (XIII, 12). That is, in a very common condition of war, one found "in all times," the life of man is just about the opposite of "solitary, poor, nasty, brutish, and short."

If the natural condition of war is not essentially or necessarily a condition of active conflict, certainly not a condition in which literally every individual is in actual conflict with every other individual, and is a condition in which humans quite commonly live with some level of prosperity that comes from their industry, what does it in fact boil down to?

The likely answer would seem to be found in Hobbes's statement about the "nature" of war being in a "disposition" to fight (XIII, 8), a statement that would seem to be consistent with the characterization of the natural condition of war as an "inference made from the passions" (XIII, 10). Getting to the heart of the notion of war found in *Leviathan,* then, requires examining the passions Hobbes discusses in relation to war. Oddly, he does not name the passions he has in mind, but he does tie them closely to what he calls the "three principal causes of quarrel" "in the nature of man": "competition, diffidence, and glory" (XIII, 6). At first glance it appears that what Hobbes means is that the natural passions of man lead him to want to surpass others, lead him to be inclined to dominate as many others as possible, and lead him to seek deference, admiration, and recognition of his superiority. In other words, it looks as though the passions of man lead him, in general, to pursue power over others, and that it is the character of the passions that makes human existence a war of every man against every man. (Again, since one does not always act on one's passions, the state of war need not be a condition of actual conflict.) When Hobbes says war is a state in which nothing prevents conflict when there is a known disposition to it (XIII, 8), it looks as though he means by the latter part (the known disposition), passions of the sort I have just described.

That would be consistent with Hobbes's general reputation and with the apparent meaning of some of his most famous words, such as, "in the first place, I put for a general inclination of all mankind, a perpetual and restless desire of power after power, that ceaseth only in death" (XI, 2). But in fact *Leviathan* is not so simple. To understand the line just quoted, for example, it is necessary to read Hobbes's account of what power is and about the multiple forms it takes—an account given in the chapter immediately preceding the chapter with that famous quotation. There we see Hobbes saying very little about power as power over other humans, but we do see him writing in praise of liberality and even friendship (X, 4). It is quite surprising on the whole, given Hobbes's reputation, how rarely, if at all, in *Leviathan,* we find arguments or assertions that the acted-upon urge to surpass and dominate other humans is normal or typically human. It is not in *Leviathan* that Hobbes quotes the old saying "Man is wolf to man." For that matter, when he does quote that in the dedicatory letter to *De Cive*—his exact words are "man to man is an arrant wolf"—Hobbes specifies that *nations* are like wolves to one another but says that within nations and referring to individuals, "man to man is a kind of God."[8]

What, then, about the passions leads us into conflict with each other? To begin, the initial appearance of three passions being at play in the natural condition is misleading. The second cause, which leads to diffidence and then to the greatest amount of conflict, does not appear to be a passion at all. It is not a passion but rather a calculation of what others can probably be expected to do, that leads one to distrust them and so to take wide-ranging preventative action to stop them from taking what one has acquired. Hobbes says, "there is no way for any man to secure himself so reasonable as anticipation, that is, by force or wiles to master the persons of all men he can, so long till he see no other power great enough to endanger him" (XIII, 4). We are left with two passions, one which leads to competition and one which is connected to the human concern for reputation and esteem, and that is to say that it appears that the two passions in question drive us to try to surpass others and to seek recognition of superiority from others. So despite what I asserted just above about *Leviathan* in general, in this specific instance it nevertheless appears that the conflict-causing passions are passions for eminence of various kinds.

On examination, however, it turns out that the passions in question are not accurately described in that way. Hobbes does not in fact say that competition results from a desire to have more than others or to surpass them in some other way. Similarly, the passion connected with "glory" is not, as Hobbes describes it, a passion that leads one to actively seek recognition of one's superiority; he does not say it is a passion that sends one in quest of admiration.

How does conflict begin, then? Hobbes says it this way: "From this equality of ability ariseth equality of hope in the attaining of our ends. And therefore, if any two men desire the same thing, which nevertheless they cannot both enjoy, they become enemies" (XIII, 3). It is not perfectly clear what Hobbes is referring to by "this equality of ability." He has spoken of the equality of all men in "the strength of body" and equality of all in "the faculties of the mind" (XIII, 1), and it could be that "this equality of ability" includes both.

However, it seems we can say that that the equality of the faculties of the mind is—at least—the more important of the two. Hobbes says himself that he finds "a greater equality amongst men" in the faculties of mind "than that of strength" (XIII, 2). If it is equality of ability that leads to equality of hope, then where there is greater equality of ability there will be more equality of hope. But it could be that equality of mind is the only equality of ability that leads to equality of hope. Hobbes explains the equality in faculties of mind by noting that "men will hardly believe there be many so wise as themselves," and "there is not ordinarily a greater sign of the equal distribution of anything than that every man is contented with his share" (XIII, 2).[9] Hobbes does not say that men typically believe that there are few others as strong as themselves or that "every man is contented with his share" of bodily strength. It would not make sense for him to say that since it is easy to see the bodily strength of others and to compare it to one's own, and it is easy to recognize that one might need more bodily strength for any number of reasons. To put it differently, using the greatest sign of equal distribution, it would appear that men are not equal in strength. Indeed, Hobbes speaks of "the weakest" and "the strongest," and thus acknowledges that there are real differences in strength. Equality of body is not ordinarily observed or believed in. It is something Hobbes must point out to people, using his own idiosyncratic measure which goes well beyond the strength in an individual body: "the weakest has strength enough to kill the strongest either by secret machination, or by confederacy with others" (XIII, 1). It is not plausible to say that one man typically gets into competition with another in the literal belief that few other individuals are as strong as him. Competition has another origin.

The beginning of competition is in a belief each individual has about his own mind and his own wisdom. No one believes there can be many as wise as himself, and so each believes in his ability to attain what he desires (and he may desire the same thing another desires). The beginning of competition, in other words, is an opinion one has of the qualities of one's own mind. And the beginning of the conflict with others over one's reputation is a sign of undervalue that another person shows for one or for one's family, friends, nation, and so on. At the risk of boiling it down too far, when I want another to value me at the rate at which I value myself, I want him to have a specific kind of thought identical to a thought I have. Certainly I can feel undervalued by another who has neither taken nor threatened action against me. I will revisit this below, but being undervalued or shown contempt can be an injury felt entirely as an assault on one's thoughts and only on one's thoughts. The two causes of conflict connected to the passions, competition and glory, begin with what can be called particular states of mind, and so it seems quite reasonable to seek an explanation of the origins of conflict in

Hobbes's teaching on the working of the mind. He begins that teaching with an account of perception.

It makes sense to begin that way since, as Hobbes notes, "there is no conception in a man's mind which hath not at first, totally or by parts, been begotten upon the organs of sense" (I, 2). Understanding perception, it turns out, is necessary not only for understanding thinking and reasoning, but for understanding the passions and even what we might now call "value judgments." That is, an appetite or desire can be directed only at something already sensed, just as a thought or conception can be only of something already sensed. And what an appetite or desire is for, one calls good, while that from which one has an aversion one calls evil or vile (VI, 7).

The cause of a sensation, Hobbes says, is the actual physical contact of a body on an organ "either immediately, as in the taste and touch, or mediately, as in seeing, hearing, and smelling" (I, 4). Sense is caused by matter in motion striking other matter, and the most basic laws of motion apply to that matter and explain imagination, memory, and dreams, which are either decaying or obscured sense—that is, slowing or weakening motion—or the product of "the agitation [or movement] of the inward parts of man's body" (II, 5).

Hobbes makes clear that because of how we perceive, it is possible to perceive, and therefore to think of or conceive, only what is material and finite (III, 12). As he fills out the picture of the way perception works and the way that thinking follows perception he also shows, to use a term that is not Hobbes's, that any and all human thought is purely "subjective." Not only are all of my sensations dependent on my organs, and so, for example, I will sense some things more intensely and some less intensely than someone else whose organs function differently, but the train of my thoughts and of what I imagine is dependent on my specific experiences. "We have no transition from one imagination to another," Hobbes says, "if we never had the like before in our senses" (III, 2). That is because memories and thoughts work according to the laws of motion. A sensation is accompanied by, follows and is followed by, some sensations and not others in the physical sense. There is no such thing as free association since any particular sense or thought is always associated with other specific sensations or thoughts, again, in a physical, material sense.[10] My sensing mechanism may eventually be filled to its physical limit—we speak of sensory "overload"—but as long as I continue to experience new things, it is not accurate to say that my mind is a closed system. In all circumstances, however, it is a unique system, depending on the functioning of my own organs and on my unique set of experiences, and consisting of my particular set of desires, aversions, hopes, fears, and so on.

As an aside it should be noted that though my thoughts are unique, some degree of community can be attained through exchange of speeches—and no community at all could be attained without speech, Hobbes says (IV, 1). However, "all names [or words] are imposed to signify our conceptions" (IV, 24). We cannot name things directly or name things themselves, and so it follows that it would be only by coincidence that particular words could be understood in exactly the same way by any two people. Those living closely together with many similar experiences understand each other better than others. But we need think only of the misunderstandings between spouses and between parents and children, to see how rare perfect understanding is. And I will revisit below the question of how much community Hobbes thinks it is possible to attain.

Let us return to the causes of quarrel Hobbes identified. In chapter thirteen he said that competition originates in the sense each has that he is wiser than just about everyone else, and he says that that sense "is but a vain conceit" (XIII, 2). But if one's belief in one's own wisdom reflects the character flaw of vanity, Hobbes does not indicate that characters can be changed any more than the laws of motion can be changed. And in a later passage Hobbes gives a different account of the belief each has about his own wisdom. In chapter forty-four he makes a broader statement along the same lines: "no man can conceive there is any greater degree of [understanding] than that which he hath already attained to" (XLIV, 2). This statement is broader than the similar one in chapter thirteen because it means not merely that one thinks no one else is as wise as oneself, it means that at any given time one's own wisdom and understanding seem to oneself perfect or complete. In fact, the belief in the perfection of one's wisdom comes first and it then follows that insofar as one can see that another thinks differently, that other naturally seems to be less wise and so naturally does not appear worthy of deference. Yet this is not vanity as a character flaw because it is a physical impossibility to conceive of understanding greater than that which one has oneself, just as it is a physical impossibility that a man conceive "in his imagination any greater light than he hath at some time or other perceived by his outward senses" (XLIV, 2). Competition begins, to make the point plain, because of the physical impossibility of conceiving of wisdom superior to one's own, which naturally appears to oneself (but only to oneself) complete or perfect wisdom.

Before going on, it is worth considering for a moment whether what Hobbes says here is plausible. On the face of it Hobbes seems completely wrong if he is saying that we cannot imagine anything larger or of greater magnitude than anything we have already perceived. But it does not seem that he can mean, for example, that we literally cannot imagine a house larger than any house we have ever seen. After all, Hobbes says we can imagine many kinds of things that we have never seen and that do not even exist, such as centaurs (II, 4). We can imagine centaurs because we can make a kind of addition or combination in our minds of two things we have already perceived—a man and a horse—and so we can assume Hobbes thinks it possible to imagine a house or any object larger than any example of that object we have ever perceived by a similar addition or combination in our minds. But that is not necessarily the same as saying we can imagine greater wisdom or understanding than that which we have at any given moment, and that is especially so given what Hobbes seems to mean by "wisdom." It is possible to say only what he "seems" to mean because he does not define wisdom in *Leviathan;* he does not include it among "the virtues commonly called intellectual" defined in chapter eight, and he makes clear it is not the "skill of proceeding upon general and infallible rules called science," nor is it prudence or wittiness or eloquence or learnedness (XIII, 2).

I propose that a description of wisdom that fits with what Hobbes says about it is this: wisdom is the sense each individual has of the coherence of the totality of his experiences and perceptions. Wisdom defined this way is acquired by everyone merely by living, and it would not be entirely fair to say that believing that this is wisdom is to be deluded. For it is the case that at any moment one has a sense of a particular whole, the whole of one's own life. And since one's experiences and perceptions fit together, in a way simply by occurring in a sequence and with associations among them, all of one's experiences *should* seem to one to cohere. That this kind of overall sense naturally seems to one to be a kind of wisdom can be seen when thinking about when one feels least wise or least able to "make sense of it all." It is when one has a traumatic disruption to one's life, a harmful experience far from anything previously experienced, that one will ask "Why?" and in asking that question search for a new account of a coherent order to replace the one disrupted and lost.

Further, a model for wisdom along these lines can be found in a philosophic work Hobbes would find little with which he agrees, yet which he thinks *Leviathan* might resemble: Plato's *Republic* (XXXI, 41). In that work Socrates describes prisoners chained in place in a cave, able to see only shadows of artifacts, cast by a fire onto the wall of the cave.[11] Socrates acknowledges to Glaucon that his description of these prisoners is strange, yet he also claims that they are "like us," which is to say that Socrates thinks he and his companions have a similarly limited ability to perceive even the material world. This ability is so limited that when a prisoner is shown things he is not accustomed to seeing, he cannot perceive those things at all. Yet in the entire account of the prisoners, their release from bondage, and their ascent into the light of the sun, Socrates speaks of "wisdom" only once. A former prisoner who has slowly become able to look at the sun itself might recall, Socrates says, "the wisdom" back in the cave.[12] Socrates does not say that one gains in wisdom the further one gets away from the original bound condition, nor even that those who contemplate the sun have wisdom. To be sure, in this passage Socrates does not define wisdom any more than Hobbes does in *Leviathan,* but he would seem to mean that wisdom is the overall sense the prisoners have of having perceived everything there is to perceive, a sense they have because of the physical limitations on their ability to perceive, limitations so absolute that it can be said of them that they cannot conceive of more wisdom or understanding than they already have achieved.[13] All of this is to say that Plato, at least, would seem to have found plausible Hobbes's assertion about the impossibility of conceiving greater wisdom or understanding than that which one has at any moment attained.

To return now to the cause of conflict among men which is the second passion that is a cause of conflict in the nature of man, the passion concerned with glory or the degree to which another shows he values me. It seems much more difficult to explain this passion using the natural science of perception and only that natural science, but perhaps something like the reasoning about the passion leading to competition applies to this one also.

If we look closely at the explanation Hobbes gives, we can see some connection to the state of mind that leads to competition. Hobbes says, "every man looketh that his companion should value him at the same rate he sets upon himself, and upon all signs of contempt, or undervaluing, naturally endeavors as far as he dares . . . to extort a greater value from his contemners" (XIII, 5). What signs of contempt, specifically, lead to conflict? Hobbes says they will be "trifles," such as "a word, a smile, a different opinion" (XIII, 7). As I noted above, Hobbes does not say we actively seek out glory or admiration or recognition of our value, nor do we want it from everyone; it is that no one wants to be undervalued by someone close to him, his companion. But what can this possibly have to do with the explanation of perception?

Perhaps a connection can be found in the following way: the signs of contempt Hobbes names are not physically threatening. One of them is a smile. The other two—a word and an opinion—have directly to do with the understanding, with thoughts. Hobbes does not tell us exactly what is so provoking about a word (or a smile) indicating contempt, nor does he say specifically why the mere expression of an opinion opposing one of mine seems to me to be a sign of undervaluing me. None of these is a direct or physical threat, but experience tells us that the mere expression of an opinion can and often does feel like a kind of threat, a threat to the sense I have of the perfection or completeness of the wisdom I seem to myself to have, a sense I have due to the physical limits of my mind. Hobbes says here that both insulter and insulted might fight to the death, and in real life they surely might do just that. And there is a kind of logic to their doing so. To accept the insult would be to accept a negation or denial of the completeness or perfection of one's understanding and wisdom, the framework of senses and thoughts one lives in naturally and as a matter of physical necessity. Life would seem to be no longer one's own; it would seem intolerable or like no life at all. The physical workings of our minds force on us the logic of fighting when we feel undervalued by another.

This explanation of the natural condition as originating in the physical working of the mind does not make that condition easier to overcome than it would be if it originated in desires to dominate or be pre-eminent. To the contrary, it is much more difficult to see how the condition of war can ever be fully overcome in this case. We might invoke "education," but in a crucial passage Hobbes suggests that the kinds of fundamental insights that are required for breaking out of one's self-contained "thought shell" come only by accident (XLIV, 2). There certainly is evidence that Hobbes did not think the condition of war can be overcome even with an all-powerful sovereign. Most simply, it does not appear that even with such a sovereign one will typically come to think there is no need to lock one's doors and safes and so on. When you believe in the need to do that, Hobbes tells us, you think you are in a kind of state of war against whoever you fear might rob you, and in his example that includes members of your own family (XIII, 10). Nor with an all-powerful sovereign does Hobbes expect any individual to forget himself; no individual contracts away his right to try to stay alive, no matter what. What is more, if the all-powerful sovereign were sufficient to bring peace to a commonwealth, there would be no need for Hobbes to spell out so many laws of nature in chapters fourteen and fifteen of *Leviathan.* The summation of those laws is especially revealing: do not do to another anything you do not want done to yourself (XV, 35 and XIV, 5). The version of that law in the Gospels—do unto others as you would have them do unto you—will lead to peace if I can know another's mind because it is like mine.[14] Hobbes does not ask anyone to try to take into account another's state of mind. Rather, his path to peace, like his way of proving his doctrine (Introduction, 4), requires each to know only his own unique mind (see II, 1).

I want to conclude with a kind of personal reflection that might be inappropriate if Hobbes did not himself say at the end of the "Introduction" that his doctrine can be demonstrated only by means of individual self-examination.

Having made the above arguments, I am convinced that Hobbes's presentation of the natural human condition is based only on his explanation of the materialistic, mechanistic workings of our minds. Yet at the same time I find that I myself am not able to completely accept that Hobbes is right nor even able to completely accept that Hobbes is totally sincere when he makes his explanation of the mind so important. And it could well be that what makes all this so difficult to accept is not a flaw in Hobbes's argument or in his evidence but rather the pride I have, and that I suspect I share with most people, in the working of the human mind, a pride that makes me think that our minds are somehow in part divine, even though we may rarely use that term.[15] But perhaps it is precisely that particular manifestation of human pride that Hobbes is above all trying to dispel—by what exact mechanism is unclear—with his *Leviathan.*

ENDNOTES

1. See Cees Leijenhorst, "Sense and Nonsense about Sense: Hobbes and the Aristotelians on Sense Perception and Imagination" in *The Cambridge Companion to Hobbes's* Leviathan, ed. Patricia Springborg, (Cambridge, Cambridge University Press: 2007), 82.

2. Leijenhorst, Sense and Nonsense, 82.

3. I have used the edition of *Leviathan* edited by Edwin Curley (Indianapolis: Hackett, 1994). I have omitted page numbers and instead use only chapter number in Roman numerals (or chapter name) followed by paragraph number.

4. Without explanation, Hobbes translates the Latin *nosce teipsum* as "read thyself" rather than "know thyself." His idiosyncratic translation is likely meant to be consistent with the idiosyncratic explanation of the Apollonic exhortation he gives in the Introduction. Much of *Leviathan* makes arguments against Cicero's explanation of "*nosce te*" found in *Tusculan Disputations* I, 52 and against much else in Book I of that work. There Cicero compares himself to Aristotle, argues that death is not something wretched, says that "know thyself" must mean "know your soul," and notes that "unless we are unable to understand what something is that we have never seen, certainly we can form a conception of god himself and of the divine soul liberated from the body." (Hobbes pairs Cicero and his clear opponent Aristotle several times in *Leviathan.* See IV, 13; XXI, 9; XXVI, 1; XLVI, 11 and 37.) Prof. Jospeh McAlhany of Carthage College assisted me with the *Tusculan Disputations.*

5. Hobbes does seem to say that at times one can know "the thoughts and passions of all other men" by looking into oneself. However, Hobbes goes on to say that it is all but impossible to know in others "the objects of the passions, which are the things desired, feared, hoped & c" (Introduction, 3), and the latter knowledge is the more important of the two. For a contrary explanation of this passage, see C. B. Macpherson *The Political Theory of Possessive Individualism: Hobbes to Locke* (Oxford: Oxford University Press, 1962) 30.

6. Scholars note the topic of "reading" oneself and others in the Introduction even more rarely than they note the fact that the first three chapters of *Leviathan* are on the inner human life. For example, *The Cambridge Companion to Hobbes's* Leviathan contains only four references to the Introduction (pp. 33, 61, 63, and 174), and three of those are to the same sentence (on what a Leviathan is). None of the four is to anything past the first half of the first paragraph. In twenty-two essays and almost

five hundred pages on *Leviathan*, no mention is made of the topic Hobbes treats at greatest length in the Introduction.

7. Michael Oakeshott, in his Introduction to his edition of *Leviathan*, says that for Hobbes "The nature of man is the predicament of mankind," and while that sounds like a summation of Hobbes on the natural condition, in fact Oakeshott says that leading into his overview of Hobbes on the inner human life. See Michael Oakeshott, ed., *Leviathan* (Oxford: Basil Blackwell, 1960), xxx.

8. *Man and Citizen*, Bernard Gert, ed., (New York: Humanities Press, 1978), 89.

9. At first it appears that Hobbes is saying that the minds of all men are equal because they all have equal prudence. But on examination it turns out that only those who "equally apply themselves" in an equal amount of time have equal prudence (XIII, 2). Equality of mind is equality of wisdom, and that equality is proven as indicated in the text.

10. This can make Hobbes sound extremely simplistic, but one need only look at his example of a question posed "in a discourse of our present civil war," namely, "what was the value of a Roman penny?" to see that he knows that multiple, complicated thoughts can follow one another extremely rapidly. As Hobbes says, "thought is quick" (III, 3).

11. See Plato, *Republic* 514a–515a.

12. *Republic* 516c.

13. The allegory of the cave is by no means an account of the human condition Hobbes would fully accept, perhaps above all because the prisoners seem able to have the same understanding of things. Hobbes thinks a human being has the "character of the solitary," not because he is disinclined to live with others but because he "lives in the world of his own sensations and imaginations" (see Oakeshott, xxxiii).

14. See *Matthew* 7:12 and *Luke* 6:31.

15. See, for example, Aristotle, *Nicomachean Ethics,* 1177b30–31 and Cicero, *Tusculan Disputations*, I, 65.

Chapter 10

Hobbes and Aristotle: Science and Politics

Leah Bradshaw

In much political theory discourse in the last decade or so, the state has become unfashionable, and theories of globalization, "one world" and cosmopolitanism have predominated. The commitment behind the globalization agenda is formulated in different ways. Theorists of trans-political loyalties may point to the universality of human rights that ought to trump any state or cultural loyalty; more recently, there is a body of "cosmopolitan" theory that calls for an identity that transcends the political, and embraces all human beings. Despite the variation, behind these theories is a common commitment to autonomy as the first and primary goal of human beings.

Competing with the "one world" theorists are their detractors, who do not defend the sovereign state as an alternative, but rather the integrity of distinct culture, the particularity of place and space. It seems sometimes that we are locked into a vacillation between universal claims (for human rights, or cosmopolitan reason) and what Jay Bernstein has termed "provincialisms." The reification of universalist claims according to Bernstein has led to "difficulties and disasters whose correction involved the acknowledgment of new provincialisms, such as the local laws of a natural habitat or ecological system, or the role of local and nonrationalized paradigms in scientific research; the claims of particular groups and communities against the bias of neutral equality; the interpretive perspective from which 'objective' features of social life come into prominence, and so on."[1] In Bernstein's view, the push toward universalism actually has propagated the reverse strain toward these provincialisms, thus pointing perhaps to the inherent flaws of a universalist/cosmopolitan agenda.

In both the cosmopolitan/universalist picture, and its postmodern refraction into splintered provincialisms, the political seems to have disappeared. In this paper, I want to investigate what I take to be the most powerful philosophical foundation of autonomy in the work of Thomas Hobbes. Hobbes founds his new political science on a premise that all human beings are the autonomous bearers of natural rights to defend themselves.[2] Hobbes makes claims for an infallible, universal, and scientific basis for comprehending human motivations, and for organizing human beings. The paper subsequently turns to Aristotle as the defender of politics. Aristotle gives us a view of what is "natural" for human beings, but one that does not lead to any universal political prescriptions. Neither does it lapse into a relativism of unmeasurable identity. Aristotle's recommendations for organizing political communities are varied and pragmatic, attuned to cultural, geographical, and historical considerations. There is no single "science" of government for Aristotle, though certainly Aristotle has a sense of what human beings are fitted for. There are no "rights" in Aristotle's political thought, but there are ethical considerations on how people ought to treat one another when they inhabit together a particular kind of life. The paper will try to show how Aristotle's political science rests upon an alternative philosophical foundation from Hobbes's, and that Aristotle provides a persuasive argument for how we can recognize the real and substantive variety among communities without abandoning the possibility of political integrity. I hope to make an argument that Aristotle's political thought is a good basis from which to conceive an alternative to rights and autonomy universalism, and is a way of thinking about politics that is open to cultural diversity without necessarily sinking into relativism. I realize of course that there is a massive body of political theory that seeks to redress the problems of liberalism by calling on Aristotle's "communitarian" sentiments, but I am more interested in this paper in exploring the underlying epistemological differences between Aristotle and Hobbes, and how these might be central to their differing political schemas.

Knowledge of any kind for Hobbes has to be tracked back to sensation. Hobbes tells us at the beginning of *Leviathan* that all our thoughts can be traced back to sense response, and sensations always arise in us as a consequence of collision and friction. Something impacts on our nerves "and other string and membranes of the body, continued inwards to the brain and the heart," causing a "resistance or a counterpressure."[3] We exist in the world as a bundle of sensations, and our most formative experiences are ones of conflict and resistance, to the elements, to other human beings and to the world in general. Hostility is our primal mode of being. Prudence, Hobbes defines as the simple adding up of experiences that cautions us against dangers and allows us to measure more accurately how to preserve ourselves. This is something we share with all sentient beings; it is not peculiarly human. What is unique to human beings is the capacity for speech.

Speech has two purposes, according to Hobbes. One is to facilitate memory, so that by having a bank of words, we can recall past experiences better, which in turn, will allow us better to anticipate the future. A second advantage of speech is its ability to put a name to a desire or fear, or some other passion. [Hobbes, Lev.13]. Importantly, speech and words are the servants of the passions. They are clarifiers and organizers of primal sensations, and the better they are at performing these tasks, the more accurate they are. Indeed, Hobbes defines science as the "right definition of names." [Hobbes, Lev.15]. Hobbes expends much effort explaining how it is that words can deceive. "Natural sense and imagination are not subject to absurdity. Nature itself cannot err" [Hobbes, Lev.15]. But as human beings develop a greater sophistication in language, their capacities for both wisdom and folly abound. With letters, a man can become "excellently wise" or "excellently foolish" [Hobbes, Lev.15]. Words are the "money of fools" and Hobbes warns us against taking countenance from such word wizards as Aristotle or Cicero or Thomas Aquinus [Hobbes, Lev.15]. The Greeks erred, Hobbes assures, in identifying reason and speech in the common term of logos. For reason and speech are clearly distinct in Hobbes's account. Reason is a response to the senses, is prudential concern for self-preservation and inherently suspicious of external threat; speech is the collection of names that we acquire to facilitate our getting by in the world, and since we are creatures inclined to fear, diffidence, and sometimes glory, we will use words to manipulate and deceive. Speech alone cannot be trusted to give an accurate account of things, for "one man calleth wisdom what another calleth fear, and one cruelty another justice, one prodigality what another magnanimity, one gravity what another stupidity, etc" [Hobbes, Lev.17].

Hobbes's distrust of words, and his reliance upon sensation as the more trustworthy basis of human experience, resonates throughout modernity, as Charles Taylor indicates in his study of the modern turn toward "inwardness." According to Taylor, there is a new "centrality of constructed orders and artifacts in mental and moral life."[4] Taylor calls this turn a "poietic" one, in which "words are ultimately given their meaning arbitrarily through definitions which attach them to certain things or ideas" (Taylor, 197). Words become fanciful, and expressions of individual identity, and we use them to project the image that we want to present in the world. According to Taylor, we can see the reverberations of this new poiesis in both Hobbes and Locke, in Descartes, and later in the eighteenth- and nineteenth-century philosophers and novelists. Speech and words may be capable of unleashing introspective creativity on the part of individuals, but they cannot serve in any way as the glue that binds the commonwealth.

For Hobbes, we have a situation where we all begin from universal experience of sensation (rooted in collision with external forces), but we have varying accounts in speech of our depictions of reality, because we manipulate words to enhance our own advantage. Nature does not lie, but people do. How do we address this conundrum? Hobbes says that we can get out of this predicament once we understand that reason properly employed is a matter of industry, not prudence: "first, in apt imposing of names and secondly, by getting a good and orderly method in proceeding from the elements, which are names, to assertions made by connection of one of them to another, and so to syllogisms, which are the connection of one assertion to another, till we come to a knowledge of all the consequences of names appertaining to the subject at hand" [Hobbes, Lev.21]. Once we set ourselves to this task, we are doing science. More simply put, Hobbes says that "science is the knowledge of consequences" [Hobbes, Lev.21]. Enlightenment for human beings is "perspicuous words." On the road to enlightenment, "reason is the pace; increase of science the way; and benefit of mankind, the end" [Hobbes, Lev.22]. Scientific knowledge, it is important to note for Hobbes, is described as "conditional" [Hobbes, Lev.40], not absolute, and that is because science is attendant on the consequences it produces. If the benefit of mankind is the end, and science is the way, then we need to know more about what the benefit of mankind is.

To Hobbes, the benefit of mankind is simple to understand, if harder to achieve. Happiness for us consists in "continual success in obtaining those things which a man from time to time desireth, that is to say, continual prospering" [Hobbes, Lev.30]. There is no stasis of happiness, no "perpetual tranquility," because "life itself is but motion and can never be without desire, nor without fear no more than without sense" [Hobbes, Lev.30]. Science applied in light of considerations of the benefit for mankind leads us directly to consider political arrangements, for if happiness consists in "continual prospering," in the quenching of desires, and the maintaining of motion, we need to secure the foundations for these pursuits.

Hobbes's prescriptions for political commonwealth are well known. People must forfeit their natural liberty as autonomous agents in the natural condition, for the security of the state. All must submit equally to absolute authority in exchange for peace. Given Hobbes's assumptions that human beings are first and foremost creatures of sensation and passion, that they are secondarily creatures of calculative prudence (able to add up experiences to predict what is in their best interest), and that they are thirdly, capable of speech acts that can be wildly deceptive, the only way that people can live together in peace is by relinquishing their natural liberty in a contractual act that will bind them by universal law, discovered by the pace of reason, pursued by the way of science, and intended for the benefit of all mankind.

Let us be clear about the universalism of Hobbes's political science. It requires that we accept a number of assumptions. Human beings are first and foremost appetitive creatures whose primal response to the world is a sensual, not a reasoned, one, and whose appetites always incline toward the maximization of individual power, either because of a desire to dominate, or from a weaker desire simply to preserve one's own territory and body. The variability in ethical claims that one finds in the world is due to the ingenuity of human beings in employing words to advance their own self-interests. Political communities, if they are to be secure, have to be built upon the transference of right on the part of autonomous, desire-maximizing individuals, who are willing to trade absolute liberty for peace. This is a formula that Hobbes claims provides the foundation for a secure commonwealth, in all cir cumstances, in all cultures. In the law of nature consists "the fountain and the original of justice" [Hobbes, Lev.71]. The science of natural law, as defined by Hobbes is "the true and only moral philosophy" [Hobbes, Lev.79].

Just how reliable are Hobbes's claims to have founded a "true and only moral philosophy," and what is the real impact in the world of his findings? Hannah Arendt gives a different twist on Hobbes when she claims that Hobbes is "the only great thinker who ever attempted to derive public good from private interest and who, for the sake of private good, conceived and outlined a Commonwealth whose basis and ultimate end is accumulation of power."[5] Arendt argues, following Hobbes's logic, that if reason is nothing more than "reckoning" of consequences, then the very idea of a free and autonomous will is impossible. Without free will, it is difficult to construct any notion of political responsibility. Human beings are driven by the twin propellers of power and insecurity.

Arendt makes the case that Hobbes is a profoundly unpolitical thinker in the sense that he provides no grounds upon which we could imagine people sustaining a political community or a commonwealth on grounds other than a temporary utility. As Hobbes tells us in *Leviathan*, we cannot expect that any man would defend his country if he were in a place of mortal danger, and that would include if he were taken prisoner by another commonwealth. Those who live outside the protection of the commonwealth (strangers, slaves, refugees) have no moral or political obligations whatsoever, and can be expected to engage in any acts they deem necessary to preserve themselves. "Thus membership in any form of political community is for Hobbes a temporary and limited affair which essentially does not change the solitary and private character of the individual (who has "no pleasure, but on the contrary a great deale of griefe in keeping company, where there is no power to overawe them all") or create permanent bonds between him and his fellow men" (Arendt, 140).

The claims that Hobbes makes for his "true and only" moral philosophy, Arendt regards as a mask for Hobbes's true motives in justifying the unlimited acquisition of power. Hobbes presents his ontology "as though he started from a realistic insight into man, a being that desires "power after power," and as though he proceeded from this insight to a plan for a body politic best fitted for a power-thirsty animal" (Arendt, 140–141). In fact, says Arendt, things work the other way around. Hobbes conceives this new order "for the benefit of the new bourgeois society as it emerged in the seventeenth century and this picture of man is a sketch for the new type of Man who would fit into it. The Commonwealth is based on the delegation of power, and not of rights" (Arendt, 141).

Arendt emphasizes two important consequences of Hobbes's construction: driving individuals into an atomistic and competitive sense of themselves, and the state's thrust toward unlimited expansion. If security requires that people trade natural liberty for peace, then "the individual, to whom public and official life manifests itself in the guise of necessity, acquires a new and increased interest in his personal and private fate. Excluded from participation in the management of public affairs that involve all citizens, the individual loses his rightful place in society and his natural connection with his fellow-men" (Arendt, 141). Hobbes's conception of human beings as solipsistic, fearful, and power-driven coupled with his insistence that words are but the embellishments of radical self-interest, Arendt sees as intrinsically connected to the emergence of the modern "poiesis" identified by Charles Taylor. The novel as art form triumphs in this new firmament, replacing the more public and political art form of the theatre. Drawing on the example of Balzac, Arendt wrote that "Balzac showed the full range of the new genre and even presented human passions as man's fate, containing neither virtue nor vice, neither reason nor free will. Only the novel in its full maturity, having interpreted and reinterpreted the entire scale of human matters, could preach the new gospel of infatuation with one's own fate that played such a great role among nineteenth century intellectuals" (Arendt, footnote, 141).

While the thoughtful turn to the inwardness of sentiment and self-absorption, the state turns to ever increasing power. Again following Hobbes's own logic encapsulated in his famous injunction that by nature human beings pursue "power after power," a process that ceases only in death, Arendt argues that a community based on the harnessing of power cannot stand still. "Only by acquiring more power can it guarantee the status quo; only by constantly extending its authority and only through the process of power accumulation can it remain stable" (Arendt, 142). Hobbes's commonwealth has to be imperialistic. Hobbes's "true and only moral philosophy," from this standpoint, appears as a justification for universal tyranny. Hobbes's political philosophy, grounded in the senses and the passions, disdainful of the traditional coupling of reason and speech, leads to the unlimited expansion of power in a manner that will destroy any political particularity, and effectively neutralize any resistance by turning the public citizen into a private producer or consumer of poetic reverie. "A Commonwealth based on the accumulated and monopolized power of all its individual members necessarily leaves each person powerless, deprived of his natural and human capacities. It leaves him degraded into a cog in a power-accumulating machine, free to console himself with sublime thoughts about the ultimate destiny of this machine, which itself is constructed in such a way that it can devour the globe simply by following its own inherent law" (Arendt, 146).

Hobbes tells us that he is beginning from first principles, and he attempts to convince his readers that by starting with an investigation into how we know things (through senses, not reason, through passions, not virtues), he can demonstrate unequivocally that he has a foundation for politics that is scientific and universally true for all people in all time. Hannah Arendt tells us that this is a ruse, that Hobbes has constructed a brilliant ontology to justify what the ancients would have slighted as tyrannical ambition. "By Victory or Death, the Leviathan can overcome all political limitations that go with the existence of other peoples and can envelop the whole world in its tyranny" (Arendt, 146).

Science or tyranny? Which best characterizes the thought of Hobbes? Perhaps one way of deciding this is to look at an alternative construction in Hobbes's nemesis, Aristotle, to see if Aristotle provides a more comprehensive, and therefore convincing ontology. There is no "true and only moral philosophy" for Aristotle, although moral philosophy certainly hinges on some understanding of truth. As in Hobbes, Aristotle's moral and political teachings follow in an inductive manner from his "first thoughts" on how we know things. Aristotle opens the Metaphysics with the claim that "all human beings by nature stretch themselves toward knowing."[6] The first sign of this, he continues, is "our love of the senses; for even apart from their use, they are loved on their own account, and above all else, the one through the eyes" (Aristotle, Met. 980a21). Already here, we see a contestation with Hobbes, who contrarily asserts "as in sense that which is really within us is only motion, caused by the action of external objects (but in appearance, to the sight, light and colour; to the ear, sound; to the nostril, odour, etc); so when the action of the same object is continued from the eyes, ears and other organs to the heart, the real effect there is nothing but motion or endeavour, which consisteth in appetite or aversion to or from the object moving" (Hobbes, Lev. 25). Aristotle claims we love the experiences of the senses for themselves; Hobbes rejects this, claiming that no sensation can be experienced except as either an attraction or an aversion to an external motion.

Aristotle's depiction of sense as something we experience in and for itself, has parallels throughout the Metaphysics in the way that he accounts for knowledge, art, and philosophy. Unlike Hobbes, who locates all human intention in the original chaos of perpetual motion, Aristotle asserts that all human action, and all human thought and speech, incline toward purpose. It does not make sense, Aristotle tells us, to imagine the universe or ourselves as bundles of infinite reactions. "It is clear," he says, "that there is some source of things and that the causes are not infinite either in a straight line or in kind. For it is not possible for one thing to come from another infinity, either as from material (such as flesh from earth, earth from air, air from fire, and in this way without stopping) or from where the origin of motion is (such as for a human being to be moved by the air, this by the sun, the sun by strife, and for there to be no limit to this)" (Aristotle, Met. 994a).

What Aristotle insists, and what Hobbes explicitly disputes, is that there is purpose in the ordering of things, both in nature in its brutish manifestations, and in our artful constructions. To understand something is to attempt to grasp its telos. This is not easy, nor is it a precise science. Causality, for example, is multidimensional. If we want to understand the cause of something, we have to try to understand how it came into being, and how the origins are still extant in what we have before us, as for instance "the bronze of a statue, or the silver of bowl" (Aristotle, Met.1013a20). We also may try to grasp causality by reading patterns, and this is what we do in the "gathering in speech of what it is for something to be," as for example the "two-in-one ratio or generally, numbers" (Aristotle, Met.1013a20). We want to know "whatever makes a changing thing change." And finally, causality "is meant as the end." We need to inquire about that for the sake of which something exists.

For Aristotle, as human beings we exist both in motion, and in the context of purpose toward a final end, which is motionless (the matter of death). Aristotle challenges those who would claim that all things are at rest, or the reverse, that all things are in motion (Aristotle, Met.1012b). "For if all things are at rest, the same things would always be true and the same things false, but this obviously changes . . . but if all things are in motion, nothing would be true and everything would be false but . . . this is impossible" (Aristotle, Met.1012b). Something has to move the things that move, and "the first mover is motionless" (Aristotle, Met.1012b30). Readers will recognize in these passages Aristotle's famous depiction of the "first mover," a notion that forms the bedrock of the major religions of the world as well as Aristotelian metaphysics. It is possible, and even necessary, Aristotle thinks, to account for both stasis and motion, for both truth and variability, for constancy and for change. This is what we live as human beings, and it colors our ethics and our politics. We are "beings in motion," capable of building a great variety of communities, cultures, ways of living and knowing, and yet we are at the same time beings who "stretch toward knowing" final causes and ultimate purposes. The variability of things in the world does not challenge for Aristotle the claims of teleological knowing, in the way it does for Hobbes.

Hobbes says that speech cannot be trusted to found or sustain political communities, because people simply give the names of virtues to the masks of their own advantage and interest. Because of the infinite variability in the naming of things, justice, Hobbes tells us, can exist only in the binding authority of a state that decrees what justice is. Logos, for Aristotle, as Hobbes rightly identified (yet disparaged) is the combination of reason and speech, and logos exists in this space between motion and rest. As Stephen Salkever notes, "Aristotle's claim that our ability to speak reasonably is the human need that explains and justifies the political order is not easy to grasp. The difficulty is caused by the fact that his sense of the function or place of human speech in human life differs radically from conceptions that are familiar to us. . . . the place of speech in human life is that logos makes it possible for us to discover, through argument, conjecture and narrative, the kinds of goals in terms of which we can most seriously organize our lives."[7]

It is perplexing, Aristotle recognized, that logos comes up with many variations on the organization of political communities and interpretations of justice, but it is not the case that the variability is infinite, or that it is not curtailed in some manner by reference to an ultimate ground. Of the political sorts of justice, Aristotle writes, "one kind is natural, and another is conventional." [8] What is naturally just "has the same power everywhere, and is not affected by whether it seems to people or not." What is conventionally just is encapsulated in the specific practices and laws of particular communities, "such as being released for a ransom of one mina, or sacrificing a goat but not two sheep" (Aristotle, N.E. 1134b20–30). There are some people, Aristotle warns, who take all of justice to be conventional because of the variability of practices and morals, and because they believe that if justice were real, it would be everywhere and always the same, just like fire that burns the same whether in Athens or in Persia. "But this is not the way it is," Aristotle says, "though it is so in a certain sense, and while among the gods, no doubt, nothing changes at all, among us there is something that is by nature even though everything is changeable" (Aristotle, N.E. 1134b20–30).

So what is justice, by nature? Aristotle does not give us a universal prescription for this, because it is impossible, but natural justice is the practice of virtue with a view to human happiness. Understanding justice simply, requires an investigation into causality: what our origins are, and the ends for which we are fitted. *The Nicomachean Ethics* and the *Politics* read together give us a good idea of what "natural" justice is, and it includes the exercise of both intellectual and moral virtue. The intellectual virtues of wisdom and prudence are those that anchor us to the things that are everywhere and always the same. The moral virtues are always exercised in relation to others in specific contexts, and therefore their particular manifestations will vary, but they are practiced with a view to realizing a good. No one can lead a purely intellectual life, since we are rooted as human beings in the struggle for survival and in particular contexts of cultural specificity. Unlike Hobbes, however, who grounds his new political science in the mortal struggle, Aristotle advises that "as far as possible, one ought to be immortal and to do all things with a view toward living in accord with the most powerful thing in oneself, for even if it is small in bulk, it rises much more above everything else in power and worth" (Aristotle, N.E. 117b30).

Aristotle's political science, unlike Hobbes's political science, does not lead to one monolithic prescription for all orders. Aristotle recognizes the plurality of moral doctrines, and the plurality of political communities, that necessarily characterize human existence. Strangely, perhaps, Aristotle's "final cause," and his commitment to the "first mover" that underscores flux and change, opens him to a far greater toleration of diversity than Hobbes. In the *Politics*, he writes that "the variety of regimes—how many there are and in how many ways they are combined—should not be overlooked. And it is with this same prudence that one should try to see what laws are best and what are fitting for each of the regimes. For laws should be enacted—and all are in fact enacted—with a view to the regimes, and not regimes with a view to the laws."[9]

Adapting people for what they are naturally fitted for is always a struggle, and it is a political task. Aristotle argued that man is by nature a political being, a being fit for life in the polis, and this is because people have the deliberative faculty that allows them to arrange their independent relations in accordance with a view of justice. Human beings have basic needs (for sex, food, shelter), they have passions (like anger, love, jealously, pity, and so on), and they have reason which allows them to some extent to moderate and direct the appetites and passions. To reduce human beings to their most primal appetites, or to their most base passions, as Hobbes does, surely is to give us a false picture of what human beings are in all their complexity. Aristotle does not abandon all hope for human beings living together according to some reasonable plan, laid out in speech. Precisely because we have the capacity for deliberation, Aristotle argues that we are capable of being both the best and worst of creatures. Deliberating in the right way about the right things is always a hard task, and it requires not just the faculty of thought and the facility for speech, but the whole edifice of custom, habituation, culture and law to form it properly. Anyone who lives outside the bounds of the polis, Aristotle says, is one who acts like a beast or a god, and both are inhuman paths. Aristotle specifically identifies the nonpolitical creature as one who has an inclination toward war. "He who is without a city through nature rather than chance is either a mean sort or superior to man; he is 'without clan, without law, without hearth,' like the person reproved by Homer, for the one who is such by nature has by this fact a desire for war, as if he were an isolated piece in a game of chess" (Aristotle, Politics, 1253). The contrast with Hobbes could not be drawn more sharply. For Hobbes, people by nature everywhere are as isolated pieces in a game of chess, and political covenant is merely a mechanism for getting us out "that miserable condition of war which is necessarily consequent upon the natural passions of man" (Hobbes, Lev. 85).

There is no single "moral science" that can be ascribed to all forms of living together, no single set of laws that can be proscribed, but there are nonetheless reasonable grounds for distinguishing between better and worse sorts of political arrangements. The better ones are always rooted in some bond of friendship. In a friendship between individuals, there is a sharing of virtue, and one cares for one's friend as if she were another self. Between genuine friends, Aristotle tells us, there is no need for justice. Political communities, though they can never approximate the bonds of true friends, ought nonetheless to bear friendship in mind as the standard against which ties of belonging are measured. Political communities cannot survive without a sense of "concord" among the citizens, and this leads them to care in some way for each other, despite their differences in needs, abilities and aspirations. A citizen by definition is one who participates in the offices of the state and can take his turn ruling and being ruled. Citizenship presupposes a kind of equality that is defined by the boundaries of the state. The friendship of politics, Aristotle says is not merely a "sharing of belief," because strangers (and cosmopolitans) can share the same beliefs or principles. A city is said to be in concord "when the [citizens] agree about what is advantageous, make the same decisions, and act on their common resolution" (Aristotle, N.E., 1167a25). Aristotle devotes a whole book of the *Politics* to a discussion of "like-mindedness" that he regards as the binding glue of well-ordered political communities, and he gives us examples of what he means by this. They include agreement on whether offices should be elected, or if we ought to forge an alliance with another country, or who should rule (Aristotle, Politics, 1167a30–2). As one commentator on the *Politics* has observed, "notice that all these are points of considerable generality; citizens can agree to hold elections, but vote for different candidates; they can favor an alliance, but disagree about its terms; they can choose a leader without knowing beforehand what his commands will be. Evidently, like-mindedness is consensus among citizens about the fundamental terms of their cooperation. Disagree as they may about details and particulars, their relationship is not poisoned by hostility, suspicion and bitterness. Their disagreements take place against a broader background of trust and consensus." [10]

How can Aristotle's political science illuminate contemporary understandings of politics? We seem to slip easily into binary opposites in our thinking about politics on a global scale. Some of us are cosmopolitans, insisting that there must be one true account of human knowing that underscores just political institutions, and that way of knowing starts with the assumption that human beings are autonomous bearers of rights. Arash Abizadeh makes the case for a world without borders based on this commitment. "I assume that the core value of both liberalism and democratic theory is personal autonomy, and that freedom is valuable precisely insofar as it serves autonomy."[11] Martha Nussbaum in her now famous essay on patriotism and cosmopolitanism, first published in the *Boston Review* in 1994, defends the cosmopolitan ideal of autonomy, even though she concedes that "becoming a citizen of the world is often a lonely business . . . a kind of exile—from the comfort of local truths, from the warm nesting feeling of patriotism, from the absorbing pride in oneself and one's own." [12] Through a careful consideration of Hobbes, and the early modern formulation of human autonomy, we may be wary of the projects of "open borders" and cosmopolitan identity. One world politics rests on one world science, and that science is based on a questionable foundation.

What about the postmodern critics of hegemonic rights theory, advancing the cause of what Bernstein called the "provincialisms"? Stanley Fish, one of the most strenuous opponents to universalism of any kind, argued recently that all interests are special interests. All people's interests "proceed from some contestable point of view—and none is 'generally human.'" "An identity politics voter says, in effect, I don't care what view he holds, or even what bad things he may have done, or what lack of ability he may display; he's my brother, or he's my kinsman, or he's my landsman, or he comes from my neighbourhood, or he's a Southerner, or (and here the tribe is really big) my country right or wrong." Identity politics "is particularist whereas liberalism is universalist."[13] In this overly simple binary construction that Fish has set up, we have the possibility of universalist autonomy, justified by rights and cosmopolitan humanity, on the one side, or we have the particularist defense of blood, kin, country and identity on the other side. Aristotle avoids these binaries. We can talk meaningfully about some goods that can be ascribed to all human beings, and those include the "stretching toward knowledge," as well as the moral virtues that enhance our living together in communities. Any well-ordered community has ways of recognizing courage, moderation, and justice in its people. Cruelty, violence, and slavishness are disparaged. Well-ordered communities require lawfulness, and this in turn requires that relations of commodious living transcend kinship ties and are bound to some reasoned articulation of association grounded in speech. Trust and consensus are of course Aristotle's background conditions for genuine politics, insofar as politics requires citizens who are capable of both ruling and being ruled. Aristotle's view of human beings as "embedded" creatures, meshed between a cosmic order that is not of our creation and political and social constructions that are the products of our own reason, action and speech, leads him to defend a plurality of political communities and ways of life that struggle to harmonize this in-between state. Political communities are not like burning bushes: they do not observe exactly the same laws everywhere, but the better ones aim for justice. Aristotle's metaphysics and his politics are radically different from Hobbes's, for in Hobbes's commonwealth, justice is an artifact put together by fundamentally autonomous and solitary creatures. Aristotle's prescriptions for politics are grounded in reason and speech, bracketed by acknowledgement of a "prime mover," open to the diversity of ways of life and plurality of moral practices, and cemented by friendship.

ENDNOTES

1. Jay Bernstein, "The Causality of Fate," Maurizio Passerin d'Entreves and Seyla Benhabib, *Habermas and the Unfinished Project of Modernity* (Cambridge, Massachusetts, MIT Press, 1997) 264.

2. Hobbes is not the only modern thinker to advance a universalist ontology for grounding politics. Many theorists of cosmopolitanism draw on Kant for their foundation. Elsewhere, I have written about the differences and similarities between Hobbes and Kant with respect to the globalizing agenda, and have concluded that the differences, while significant, between their two accounts, do not mitigate the fact that the state in both cases is an instrument for the protection of individual autonomy. ("Empire and the Eclipse of Politics," David Tabachnik and Toivo Koivukoski eds.,

Empires Ancient and Modern, forthcoming Toronto, University of Toronto Press, 2009).

3. Hobbes, *Leviathan*, ed. Peter Martinich (Peterborough, Ontario, Broadview Press, 2002) 3.

4. Charles Taylor, *Sources of the Self* (Cambridge, Mass., Harvard University Press, 1989) 197.

5. Hannah Arendt, *The Origins of Totalitarianism*, (New York, Harcourt Brace and Co., 1973) 139.

6. Aristotle, *The Metaphysics*, trans. Joe Sachs (Newburyport, MA, Focus Publishing, 2002) 980a21.

7. Stephen Salkever, "Aristotle's Social Science," Carnes Lord and David K. O'Connor, eds. *Essays on the Foundations of Aristotelian Political Science* (Berkeley, Los Angeles and Oxford, University of California Press, 1991) 27.

8. Aristotle, *Nicomachean Ethics*, trans. Joe Sachs (Newburyport, MA, Focus Publishing, 2002) 1134b20–30.

9. Aristotle, *The Politics*, trans. Carnes Lord (Chicago and London, University of Chicago Press, 1984) 1289a5.

10. Richard Kraut, *Aristotle* (Oxford University Press, 2002) 468.

11. Arash Abizadeh, "Democratic Theory and Border Coercion: No Right to Unilaterally Control Your Own Borders," *Political Theory* (Vol. 36, No. 1, February 2008) 39.

12. Martha Nussbaum, "Patriotism and Cosmopolitanism," *Boston Review* (1994).

13. Stanley Fish, "When Identity Politics is Rational," *The New York Times* (February 17, 2008).

Chapter 11

From Metaphysics to Ethics and Beyond: Hobbes's Reaction to Aristotelian Essentialism*

Juhana Lemetti

During his life in philosophy Thomas Hobbes engaged in several debates. Among his sparring partners the dearest to him was not a contemporary (such as Descartes, White, Bramhall, or Wallis) but Aristotle. This article studies Hobbes's reactions to Aristotle and his followers.

First impressions would suggest that Hobbes's ridicule of Aristotle is thoroughly intended and his appreciation of the Philosopher sensationally low: "And I beleeve that scarce any thing can be more absurdly said in natural Philosophy, than that which now is called *Aristotles Metaphysiques;* nor more repugnant to Government, than much of that hee hath said in his *Politiques;* nor more ignorantly, than a great part of his *Ethiques.*"[1]

But as usually, first impressions are deceptive. Hobbes's phrasing is certainly impressive, but if the claims he makes are left unspecified, they remain merely words. To give Hobbes's claim more than rhetorical acknowledgement, we need to address three issues: the sources, the target, and the content of Hobbes's criticism.[2] After drafting the basic idea of the paper I shall move to the tricky questions of the sources and the target of Hobbes's criticism. I shall then analyze the different dimensions of Hobbes's criticism and their interdependence. The paper closes with the suggestion that Hobbes's reaction to Aristotle and Aristotelianism is not absolutely antithetical, but an attempt to articulate a pluralistic view of human nature and its purposes.

Hobbes's stance to Aristotelianism and especially to Aristotle is varied,[3] and Sorell has provided a helpful overview.[4] Hobbes's attitude to Aristotle and Aristotelianism can be characterized as opposing (anti), in favor (pro), (inter)dependent, or neutral (un). Hobbes was and remained anti-Aristotelian on certain issues, but when his own thinking (in particular, his political theory) matured on certain issues he became un-Aristotelian. Equally, Hobbes was dependent on Aristotle and Aristotelianism not only as a convenient target, but more substantially, that is to say, he agreed with Aristotle on certain matters (e.g. psychology), and even valued some of the works of Aristotle.[5] Even so, all Hobbes's discussions on Aristotle and Aristotelianism, as Paganini correctly stresses, share a common feature, the critique of essentialism.[6] To be precise, philosophy, partly advanced but certainly inspired by Aristotle, understands the nature and content of metaphysics wrongly, and this has theoretical as well as practical consequences.

SOURCES AND TARGET OF CRITICISM

The basic question when studying Hobbes's criticism of Aristotle and Aristotelianism is what texts and thinkers were formative in Hobbes's reaction to Aristotle and the Aristotelian tradition. However, answering the question is anything but simple. As a *novatore,* Hobbes's statements are often sweeping and, by modern standards, the explicit references are brief and vague. The evidence outside his works is notoriously scarce,[7] and at times he simply misquotes.[8] Nevertheless, some advance has and can be made.

First, it is a fair assumption that Hobbes knew most of the works of Aristotle. His education in Oxford taught him the basics of Aristotle's philosophy, and although he later belittled Oxford training, it does not change the fact that he received an education that was basically Aristotelian.[9] Equally, his works, despite their scarcity, demonstrate familiarity with Aristotle. Hobbes's acquaintance is perhaps most clear in *De motu,* a lengthy commentary on Thomas White's *De mundo dialogi tres* (1642) composed in 1643, but published only in 1973.[10] *De Motu* is not the first of Hobbes's works that comments on Aristotle and Aristotelianism but it is extensive, written in a time that is crucial in Hobbes's career, and because it is a commentary he is more precise about sources.[11] Though "octo libris physicae *Acroaseos*" and "libris quos postea quidam inscripserunt τῶν μετὰ τὰ φύσικά"[12] receive most attention, there are also passages which show that Hobbes was at home with *Categoriae, De Interpretatione, De Caelo,* and *De Generatione et Corruptione.*[13] With the help of modern scholarship,[14] the work provides a peephole into Hobbes's absorption of Aristotle and Aristotelianism. The fulcrum of the critique is the nature of metaphysics and the analysis of concepts *ens, esse,* as well as their Greek precursors, τὸ ὄν, τὸ εἶναι.[15] In brief, though the discussion in his later works might be more colorful, *De Motu* is not only relevant if we want to understand the development of Hobbes's thinking, but decisive for the discussion because it validates what is shown in his other works: a detailed knowledge of Aristotle.

The question of Aristotelianism is different. The recent historiographical trend which emphasizes that Scholastic schools and Aristotelianism were not uniform and homogenous may well be correct,[16] but Hobbes and other early modern thinkers often refuse to appreciate the variety, nuances, and achievements of Scholasticism.[17] This is clearly evident in the following excerpt:

> For the first Rector of the University of Paris, as I have read somewhere, was Peter Lombard [ca. 1100–1161], who first brought in them the learning called School-divinity; and was seconded by John Scot of Duns [1266–1308], who lived in, or near the same time; whom any ingenious reader, not knowing what was the design, would judge to have been two of the most egregious blockheads in the world, so obscure and senseless are their writings. And from these the schoolmen that succeeded, learnt the trick of imposing what they list upon their readers, and declining the force of true reason by verbal forks; I mean, distinctions that signify nothing, but serve only to astonish the multitude of ignorant men.[18]

Though in some ways misguided, there is no question about the rhetorical splendor of these and similar comments. However, they are of little use when trying to discern how Hobbes absorbed Aristotelianism. Much is shadowy, though some suggestions can be made.[19]

The form of Aristotelianism Hobbes found particularly repugnant was the late scholastic Aristotelianism, but in his criticism he is indebted to some degree to Renaissance Aristotelianism.[20] This conclusion should not be a surprise for Hobbes lived, worked, and was educated in an environment that was not only Aristotelian, but distinctively humanist.[21] One consequence of this is obvious. Medieval thinkers were incompetent in classical languages and this caused an incongruity between their views and Aristotle's original ideas.

Although the details are not entirely known Hobbes's primary target is clearly a combination of selected Aristotelian doctrines and their adaptation to Catholic dogma. In his own words, vain philosophy or spiritual darkness derives "partly from Aristotle, partly from Blindnesse of understanding."[22] In Hobbes's view, Aristotle is not innocent of "blindness," but the main emphasis is primarily on his ignorant and incompetent followers.[23]

Texts are one source, but equally interesting are people. As noted, Hobbes's scorn of Aristotle and Aristotelianism was not uncommon. During his stay in Paris in the 1640s anti-Aristotelian intellectuals surrounded Hobbes,[24] but for reasons of space I shall limit my discussion to the most likely source of influence, Pierre Gassendi (1592–1655).[25] Even here there is not a great deal of direct evidence. The surviving letters to Gassendi are rather formal and the same can be said of most of Hobbes's works.[26] Only on one occasion,

Hobbes does take a position, and that is when he contrasts himself with the Epicureanism of Gassendi and Kenelm Digby (1603–1655).[27]

A more fruitful source is the correspondence between Sir Charles Cavendish (1595?–1654) and John Pell (1611–1685), where Hobbes and Gassendi are mentioned frequently and together.[28] Another source is provided by the similar objections both Hobbes and Gassendi made about Descartes's *Meditationes de Prima Philosophia* (1641) which demonstrate more than coincidental intellectual orientation.[29] It is also significant that Hobbes had read some of Gassendi (e.g. *Disquisitio Metaphysica* (1643), but, at the present, we can only guess whether Hobbes was familiar with the milestone of early modern anti-Aristotelianism, Gassendi's *Exercitationes Paradoxicae Adversus Aristoteleos* (published anonymously 1624 in Paris). We can say, however, that a genuine influence between Hobbes and Gassendi existed and anti-Aristotelianism could be one of the issues that united them, though a detailed analysis is still to be written.

Although the evidence may appear murky and circumstantial, it is sufficient to provide a factual basis for a philosophical analysis of Hobbes's reaction to Aristotle and Aristotelianism. In doing this I shall follow Hobbes's own strategy, that is to say I shall study how the metaphysical critique reverberates with physics, psychology, ethics, and, ultimately, politics.

METAPHYSICS AND THE DOCTRINE OF SEPARATED ESSENCES

When Hobbes alleges that Aristotelian and to some extent Aristotle's metaphysics is absurd, he has two basic concerns: what is metaphysics, and how does false metaphysics impact on the study of nature.[30] Hobbes argued that "[t]he Schools" have understood the nature of metaphysics wrongly as "*supernaturall Philosophy,*" instead of the literal "*Books written, or placed after his*[Aristotle's] *naturall philosophy,*" which is the correct one. Metaphysics, Hobbes claims, provides a clear conceptual basis to the rest of philosophy and therefore is also called *philosophia prima.*[31] In *Concerning Body* Hobbes seems to agree with Aristotle about some basic concepts and in his later treatise on natural philosophy we even find him giving some credit to Aristotle.[32] Essence, form, and prime or first matter, if understood and defined correctly, are meaningful and useful concepts; even the notorious *specie* is adopted.[33] The disagreement is not then terminological or even conceptual but ontological.

In a similar vein, Hobbes's stance when discussing the question of physics is not black and white. Hobbes agrees with Aristotle what time is and seeks to correct his conception of place.[34] When turning to Aristotelian mechanics, we face the critique of essentialism:

> If you desire to know why some kind of bodies sink naturally downwards toward the Earth, and others goe naturally from it; The Schools will tell you out of Aristotle, that the bodies that sink downwards, are *Heavy;* and that this Heavinesse is it that causes them to descend: But if you ask what they mean by *Heavinesse,* they will define it to bee an endeavour to goe to the center of the Earth: so that the cause why things sink downward, is an Endeavour to be below: which is as much as to say, that bodies descend, or ascend, because they doe.[35]

Like Kepler, Galileo and others before him, Hobbes finds the idea that bodies have a natural movement according to their essence or *telos* to be absurd.

The substantial metaphysical difference between Hobbes and Aristotle and the Aristotelians is between hylomorphism and materialism. To Aristotle there are two substances, matter (gr.ὕλη) and form (gr. μορφή), whereas to Hobbes the only substance is matter. The core of the controversy is the concept and the doctrine of essence. From *Leviathan* we read: "From these Metaphysiques, which are mingled with the Scripture to make Schoole Divinity, wee are told, there be in the world certaine Essences separated from Bodies, which they call *Abstract Essences, and Substantial Formes.*"[36] The doctrine is confused.[37] Even with charitable reading, abstract essence and substantial form are "names of names," instead of things or "impressions of ideas." But even this conclusion requires the rebuttal of the sensible understanding of the very nature of the world (that it consists of bodies) as well as concessions in reasoning. In fact, concepts like separate essence are "the names of *Nothing*" and the workings of a delusory mind, and the doctrine related to them gives room for all sorts of dubious thinking in ethics and politics.[38]

ETHICS AND HUMAN NATURE[39]

If we look at how Hobbes comments on previous work on ethics and politics, his general complaint concerns lack of progression, although the main emphasis is on mistaken views.[40] Hobbes's objection to the doctrine of essences is found in his psychological and ethical criticism of Aristotle and Aristotelianism, namely in the idea that there is a universal human nature that sets the standard for the good life for each and every human being. The specific psychological misconception is that sociability is part of this universal essence. In *Leviathan,* Hobbes summarizes his view as follows:

> It is true, that certain living creatures, as Bees, and Ants, live sociably one with another, (which are therefore by *Aristotle* numbred amongst Politicall creatures;) and yet have no other direction, than their particular judgements and appetites; nor speech, whereby one of them can signifie to another, what he thinks

> expedient for the common benefit: and therefore some man may perhaps desire to know, why Man-kind cannot do the same.[41]

Immediately after this Hobbes lists six reasons for the continuous discord among humankind: competition, comparison, reason, which enables the criticism of government, speech, which enables quarrels over the meaning of good and evil, perpetual and general discontent of people, and the contractual nature of agreement. But what primarily concerns us here is the principal difference between Hobbes and Aristotle and the Aristotelians.

Two conclusions can be made from Hobbes's view. Firstly, Hobbes does not see any plausible reason to claim that sociability, understood as an innate quality of human beings, is natural. This, in my judgment, is correct. The stronger claim, however, says that the case is the opposite, namely, that there is an in-built tendency toward aggression and asocial behavior in human beings.[42] This, I believe, is not correct.

Though Hobbes often speaks of human nature, he explicitly denies that it is the cause of asocial behavior. Instead, aggression or asocial behavior does not arise from the essence of a human being, but is more likely a result of a lack of trust and is an epistemic problem. To Hobbes human nature refers to some general patterns of behavior, not to immutable intrinsic qualities. He does not deny that we have some dispositions from birth—we do, but they manifest themselves in various ways in different individuals.[43]

The difference between Hobbes's moral philosophy and that of Aristotle as well as some Aristotelians is that Hobbes refuses to commit himself to any preordained view of human nature and normative conclusions derived from this. Human beings are more or less fit for society but: "The question, which is the better man, is determinable only in the estate of government and policy, though it be mistaken for a question of nature, not only by ignorant men, that think one man's blood better than another's by nature; but also by him, whose opinions are at this day, and in these parts of greater authority than any other human writings (Aristotle)."[44] The allusion to higher blood finds a further articulation in Hobbes's political philosophy.

POLITICS AND EQUALITY

When we move on to Aristotle's political philosophy the critique falls broadly speaking into two categories: Aristotle's claim of natural inequality between human beings and certain democratic principles associated with Aristotle. Hobbes is quite frank when it comes to the first idea:

> The inequality that now is, has bin introduced by the Lawes civill. I know that *Aristotle* in the first booke of his Politiques [Chapters 3–7], for a foundation of his doctrine, maketh men by Nature, some more worthy to Command, meaning the wiser sort (such as he thought himselfe to be for his Philosophy;) others to Serve, (meaning those that had strong bodies, but were not Philosophers as he;) as if Master and Servant were not introduced by consent of men, but by difference of Wit: which is not only against reason; but also against experience.[45]

The argument has a further twist: human beings are not only equal but they are equal even when they are different from each other. This pluralistic reading of human nature and its purposes, instead of a vulgar interpretation of the *homo homini lupus* formula, is, I believe, more in accordance with Hobbes's original text.

But what about democratic principles? They seem to whirl round the following claim: freedom is possible only under the popular regime[46] which is tied to the idea that private judgment and related conceptions of good and evil override those determined by legitimate government. In addition, the idea that there is a private standard of good and evil is used to justify regicide.[47] What Hobbes wishes to show is that despite their apparent appeal to universal and indisputable truths, these ideas about good and evil lack the proper philosophical demonstration.[48]

Based on an essentialist view of human nature and ultimately the doctrine of separated essences, Aristotelian political philosophy gives credence to all sorts of religious and political fanatics and therefore is a source of the greatest evils from Hobbes's point of view: rebellion and civil war. What may escape Hobbes's readers is that this conclusion is a consequence of his radical philosophical criticism of Aristotle and Aristotelianism.

HUMAN HAPPINESS

Initially, it makes sense to suggest that Hobbes's anti-essentialism is sharpest on the question of human happiness. Whereas Aristotle has a strong, objective, and uniform conception of human nature and is a supporter of natural inequality, Hobbes is a psychological anti-essentialist who defends the natural equality of all human beings. In addition, Aristotle emphasized that the sociability of human beings is of the highest kind in the sense that they are able to communicate, reflect, and agree on the common good.[49] Lastly, Aristotle has a definite view of what constitutes happiness. As he writes in *Nicomachean Ethics:*

> Since happiness is an activity of soul in accordance with complete excellence, we must consider the nature of excellence; for perhaps we shall thus see better the nature of happiness. The true student of politics, too, is thought to have studied this above all things; for he wishes to make his fellow citizens good and obedient to the laws. [. . .] But clearly the excellence we must study is human excellence; for the good we were seeking was human good and the happiness human happiness. By human excellence we mean not that of the body but that of the soul; and happiness we also call an activity of soul.[50]

Hobbes seems to be diametrically opposed to this ethical ideal. Consider the following, yet another memorable passage from *Leviathan:*

> [T]he Felicity of this life, consisteth not in the repose of a mind satisfied. For there is no such *Finis ultimus,* (utmost ayme,) nor *Summum bonum,* (greatest Good,) as is spoken of in the Books of the old Morall Philosophers. Nor can a man any more live, whose Desires are at an end, than he, whose Senses and Imaginations are at a stand. Felicity is a continuall progresse of the desire, from one object to another; the attaining of the former, being still but the way to the later. The cause whereof is, That the object of mans desire, is not enjoy once onely, and for one instant of time; but to assure for ever, the way of his future desire. And therefore the voluntary actions, and inclinations of all men, tend, not onely to the procuring, but also to the assuring of a contented life[.][51]

This line of argument suggests that there is no single clear, distinct, and uniform conception of human happiness. The metaphysical conception of goodness is "but an idle term"[52] and yet another application of Aristotelian essentialism. Instead, Hobbes claims felicity is different for different people and different to one person at different times, but, most importantly, there seems to be no single end in human life, which is, according to the famous simile, a race.[53] What causes further puzzlement are the following two things: the conception of self-preservation and the distinction between apparent and real good. These, I suggest, constitute a conception of human happiness, which is based on a new approach to the science of politics. This new approach Hobbes himself thought to be free of linguistic manipulation of the key normative concepts of "good" and "evil" and it also acknowledges the varieties of goodness in private life. The puzzlement resides in the fact that on closer inspection the views of Hobbes and Aristotle are not antithetical.

Aristotle thought that through reason a person is able to understand his or her ideal rational essence as well as the common good and that these are in harmony. In a way, Hobbes does not deny this—his argument was that there is no *such* common good or utmost aim as the earlier, and in particular the Aristotelian, moral philosophy suggests.[54] Furthermore, Hobbes has one common good (*summum bonum*) and an utmost aim (*finis ultimus*), peace, which is in harmony with the real personal good of every rational human being, self-preservation. In *Elements of Law,* he concludes:

> For seeing we intend always our own safety and preservation, we manifestly contradict our intention, if we willingly dismiss such a one, and suffer him at once to gather the strength and be our enemy. [. . .] But since it is supposed from the equality of strength and other natural faculties of men, that no man is of might sufficient, to assure himself for any long time, of preserving himself thereby, whilst he remaineth in the state of hostility and war; reason therefore dictateth to every man for his own good, to seek after peace, as far forth as there is hope to attain the same[.][55]

Therefore even the tendency to aggression and asocial behavior is understandable and can be explained and Hobbes suggests that human beings are rational enough to find a solution.

Hobbes's moral philosophy has been characterized as subjectivism, scepticism, relativism, and contractarianism.[56] It could be also suggested that his moral philosophy is somewhat thin and, more precisely, that his notion of human happiness is a hollow one. It is true that Hobbes does not provide a conception of human happiness as thick as Aristotle and many other classical as well as modern thinkers. Nevertheless he gives a general framework within which human life may flourish.[57] Self-preservation explains, on the one hand, how common and individual good are not mutually exclusive, and that there is a common denominator between individuals. On the other hand, self-preservation is just a starting point, for it is not the only or necessarily the primary goal of human race. We also strive for what Hobbes calls "commodious living."[58] What we do not find in his political treatises is, of course, a precise account of what this commodious living consists of but this is fully in line with his anti-essentialism. In so far as the conception of the good life and happiness of an individual does not damage others or himself or pose a threat to the public peace, others should not intervene in an individual's way of life. The idea is articulated when Hobbes ponders the sphere of liberty citizens have, which, he writes in *Leviathan,* "depend on the Silence of the Law. In cases where the Soveraign has prescribed no rule, there the Subject hath the Liberty to do, or forbeare, according to his own discretion."[59]

To close: though Hobbes does not have a uniform and unequivocal conception of human happiness based on the essential qualities of human nature, he still maintains that there are certain common features in human nature, as well as a set of principles that regulate the social interaction that we are able to agree on. And though he is anything but free from the heritage of Aristotle, the novelty of Hobbes is that he does not deny that human beings normally live together and manage to do this quite well, he merely objects to the normative conclusions made from this empirical observation. Despite all the bonds of tradition, his conception of the study of human nature is closer to the modern than the classical ideal of science.

NOTES

* Earlier versions of parts of this paper were read at the Workshop on Early Modern Philosophy (University of Tartu, 2005) and the History of Philosophy Research Seminar (University of Helsinki, 2006). I wish to thank the participants in these and especially those in the ISSEI 2008 workshop *From Matter to Form* in Helsinki 2008. In addition, I am grateful to Dr Tuomo Aho, Mika Perälä MA, Dr Eva Piirimäe, Dr Kari Saastamoinen, Dr Mark Shackleton, and Dr Miira Tuominen for their help when composing the article.

The following abbreviations, editions, and conventions are used: UP = University Press; *HW* = *The Clarendon Edition of the Works of Thomas Hobbes* (various editors); *EW* = *The English Works of Thomas Hobbes*, 11 vols., ed. Sir William Molesworth (John Bohm: London, 1839–1845); *OL* = *Thomae Hobbes Malmesburiensis opera philosophica quae latine scripsit omnia*, 5 vols., ed. Sir William Molesworth (John Bohm: London, 1839–1845). *Latin Leviathan* refers to the 1668 Latin edition of *Leviathan* (*OL* III), parts of which have been translated in Thomas Hobbes, *Leviathan: With Selected Variants from the Latin edition of 1668,* ed. Edwin Curley (Indianapolis, IN: Hackett, 1994). I indicate page numbers of the translation in brackets. For example, Hobbes, *Latin Leviathan* XLVI, 498(474). In the notes section if a work is referred to without the author's name, it is Hobbes'.

1. Thomas Hobbes, *Leviathan*, 2nd revised student edition, ed. Richard Tuck (Cambridge: Cambridge UP, 1996), XLVI, 461–462.

2. The criticism is summarized in *Leviathan* XLVI, 462–473; *Latin Leviathan*, 492–502(470–476); Thomas Hobbes, *Behemoth: Or the Long Parliament*, *EW* IV, 213–218. For discussion, see Richard Serjeantson, "Hobbes, the Universities, and the History of Philosophy," in *The Philosopher in Early Modern Europe: The Nature of Contested Identity*, eds. Conal Condren, Stephen Gaukroger, and Ian Hunter (Cambridge: Cambridge UP, 2006), 113–139.

3. Hardly any full study on Hobbes in general or on some specific topic in his philosophy fails to comment on the issue in one way or another. During the course of this article I will indicate those I have found particularly helpful.

4. Tom Sorell, "Hobbes and Aristotle," in *Philosophy in the Sixteenth and Seventeenth Centuries: Conversations with Aristotle*, eds. Constance Blackwell and Sachiko Kusukawa (Aldershot-Brookfield: Ashgate, 1999), 364–379. See also Cees Leijenhorst, *The Mechanisation of Aristotelianism: The Late Aristotelian Setting of Thomas Hobbes's Natural Philosophy* (Leiden: Brill, 2002), 219–222.

5. Though Hobbes in various places (see, for example, *Leviathan,* Ch. I and Thomas Hobbes, *Six Lessons to the Savilian Professors of Mathematics*, *EW* VII,

339–340) criticizes Aristotelian psychology, above all its understanding of sense perception, he is still indebted to Aristotle's and the Aristotelian conceptual framework. For a thorough discussion, see Leijenhorst, *The Mechanisation of Aristotelianism*, Ch. 2 and "Sense and Nonsense about Sense: Hobbes and the Aristotelians on Sense Perception and Imagination," in *The Cambridge Companion to Hobbes's 'Leviathan,'* ed. Patricia Springborg (Cambridge: Cambridge UP, 2007), 82–108. Hobbes's appreciation of Aristotle's *Rhetoric* and treatises on animals was recorded by a contemporary, John Aubrey, in his *Brief Lives: Chiefly of Contemporaries*, 2 vols., ed. Andrew Clarke (Oxford: Clarendon Press, 1898), I, 357. For a discussion of the first case, see John Harwood, "Introduction," in *The Rhetorics of Thomas Hobbes and Bernard Lamy*, ed. John Harwood (Carbondale and Edwardsville, IL: Southern Illinois UP, 1986) 1–32; Leo Strauss, *The Political Philosophy Hobbes: Its Basis and Genesis*, trans. Elsa Sinclair, (Chicago, IL: University of Chicago Press, 1952), Ch. III; and Quentin Skinner, *Reason and Rhetoric in the Philosophy of Hobbes* (Cambridge: Cambridge UP, 1996), Ch. 6. The influence of Aristotle's zoological treatises on Hobbes has not been studied in detail.

6. Gianni Paganini, "Hobbes's Critique of the Doctrine Essences and its Sources," in *The Cambridge Companion to Hobbes's* Leviathan, 337–357.

7. On the reasons for this, see Noel Malcolm, "General Introduction," *HW* VI, xxi–xxvi.

8. To give a clear example: In *De Cive*, when discussing the notion of law, Hobbes refers to Aristotle's definition of law and uses *Rhetorica ad Alexandrum*, written probably by Anaximenes of Lampsacus and not by Aristotle. Hobbes's misattribution is strange for the common sources available to him, such as Laertius's *Lives* (V.22–25) or *Vita Menagiana* by an anonymous author and transmitted to posterity as part of Hesychius of Miletus's *Onomatologia*, omit the work. And as the editor of the modern collection of Aristotle's works writes: The "Spuriousness" of this work, has "never been seriously contested." See, respectively, Thomas Hobbes, *De Cive: The English edition*, ed. Howard Warrender, *HW* III, XIV.2, 169 and note 3; Jonathan Barnes, "Note to the Reader," in *The Complete Works of Aristotle*, 2 vols., ed. Jonathan Barnes (Princeton, NJ: Princeton UP, 1984), II, vii. This is the edition which is used to refer to Aristotle's works as well. For *Vita Menagiana*, see Hesychius, *Onomatologi quae supersunt*, ed. Johannes Flach (Teubner: Leibzig, 1882), 245–249. For a discussion, see Carnes Lord, "On the Early History of Aristotelian Corpus," in *The American Journal of Philology 107*, no. 2(1986), 137–161 and sources there.

9. Thomas Hobbes, "The Life of Thomas Hobbes" (Exeter: The Rota, 1979), 2–3. See also Noel Malcolm, "A Summary Biography of Hobbes," in his *Aspects of Hobbes* (Oxford UP: Oxford, 2002), 4–5 and references there.

10. Thomas Hobbes, *Critique du 'De Mundo' de Thomas White*, eds. Jean Jacquot and Harold Whitmore Jones (Paris: Vrin, 1973). I refer to this work by the original title *De Motu*, instead of the customary *Anti-White*. On White, see Beverley Southgate, *"Covetous of Truth": The Life and Work of Thomas White, 1593–1676* (Dordrecht: Kluwer, 1993).

11. The more familiar pre-*De Motu* works include: *A Briefe of the Art of Rhetoric* (1636), *Elements of Law* (1640), and *De Cive* (1642). *Briefe* is the translation of the

summation of Aristotle's *Rhetoric* that Hobbes prepared in the 1630s, mainly for educational purposes. The modern printing can be found from *The Rhetorics of Thomas Hobbes and Bernard Lamy*, 33–128. In the first two versions of his political theory, philosophical criticism of Aristotle and Aristotelianism is not prominent. The translation of the Latin propaganda pamphlet *Altera secretissima instructio* (1627) contains references to classical authors and literature, but whether there is a substantial link to Aristotle or Aristotelianism is beyond the scope of this essay. The work appears in *Reason of State, Propaganda, and the Thirty Years' War: An Unknown Translation of Thomas Hobbes*, ed. Noel Malcolm (Oxford: Oxford UP, 2007). Another piece, a prefatory note to readers in Thucydides's *History of Peloponnesian War* (1629), speaks of Aristotle as an acknowledged authority in the field of philosophy, similar to Homer in poetry, Demosthenes in eloquence, and Thucydides in history. See *The Peloponnesian War*, trans. Thomas Hobbes, ed. David Green (Chicago, IL: University of Chicago Press, 1989), xxi. The less familiar pre-*De Motu* works include three essays in the collection *Horae Subsecivae* (1620) and *A Short Tract on First Principles* (1630–1). Neither are discussed here because there is no agreement on the authorship of the works among Hobbes scholars. For discussion in the case of the former, see Noel Reynolds and Arlexe Saxonhouse, "Hobbes and the *Horae Subsecivae*," in *Three Discourses: A Critical Modern Edition of Newly Identified Work of the Young Hobbes* eds. Noel Reynolds and Arlene Saxonhouse (Chicago, IL: University of Chicago Press, 1995); Skinner, *Reason and Rhetoric in the Philosophy of Hobbes*, 236–238; and in the case of the latter: Karl Schuhmann, "Le *Short Tract*, premiére oeuvre philosophique de Hobbes," in *Hobbes Studies VIII* (1995), 3–36; Timothy Raylor, "Hobbes, Payne, and *A Short Tract on First Principles*," in *The Historical Journal 44*, no. 1 (2001), 29–58. The text of *A Short Tract* is published in *Court traité des premiers principes*, ed. Jean Bernhardt (Paris: Presses Universitaires de France, 1988). Lastly, we know that Hobbes started to draft his "Elements of Philosophy" already in the 1630s and already these documents demonstrate familiarity with Aristotle. See *De Motu*, Appendix II, 450 and 453. Both survive in the final version, see Thomas Hobbes, *Elements of Philosophy. First Part: Concerning Body*, *EW* I, II.VII.3, 94–95 and II.VIII.3, 104. Also in his first Latin optical treatise (composed ca. 1641) some familiar points are mentioned, e.g. the doctrine of visible species and Aristotle's idea of sensing as motion. See Thomas Hobbes, *Tractatus Opticus* I.10 and IV.13, in *Rivista Critica di Storia della Filosofia XVIII*, no. 2 (1963), 151 and 207. But since the primary intellectual counterpart of the work is Descartes's *La Dioptrique* (1630), I will not develop the discussion any further.

12. *De Motu* I.1, 105.

13. See, respectively, *De Motu* V.1–3, 128–131;VI.7, 141; and IX.16, 169.

14. The editors give the exact references. Hobbes's references normally contain just the name of Aristotle's works and sometimes the books in the works.

15. See esp. *De Motu*, XXVII.1, 312–314.

16. See Stephen Menn, "The Intellectual Setting," in *The Cambridge History of Seventeenth-Century Philosophy*, 2 vols., eds. Daniel Garber and Michael Ayers, (Cambridge: Cambridge UP, 1998), I, 33–86; M. W. F. Stone, "Scholastic Schools and Early Modern Philosophy," in *The Cambridge Companion to Early Modern*

Philosophy, ed. Donald Rutherford, (Cambridge: Cambridge UP, 2007), 299–327; and the review essay by Michael Edwards, "Aristotelianism, Descartes, and Hobbes," in *The Historical Journal 50*, no. 2 (2007), 449–464.

17. For some views, see Edwards, "Aristotelianism, Descartes, and Hobbes," 449.

18. *Behemoth*, 214. Cf. *Latin Leviathan* XLVI, 495–497(472–473). For a more concrete occurrence, see *De Motu* XXXVII.9, 408–409.

19. Cf. Noel Malcolm, "Thomas Hobbes and Voluntarist Theology," PhD thesis (University of Cambridge, 1983), 8–12 and Leijenhorst, *The Mechanisation of Aristotelianism*, which is most thorough. I have learned from both, but add that one source could be Porphyry whom Hobbes mentions briefly and whose *Institutiones* was adopted into the Oxford curriculum 1564/5. See *De Motu* XXVII.1; Thomas Hobbes, *Correspondence*, 2 vols., ed. Noel Malcolm, *HW* VI and VII (Oxford: Oxford UP, 1994), I, Letter 34 and 108; James McConica, "Humanism and Aristotle in Tudor Oxford," in *The English Historical Review*, 94:2 (1979), 292. On the general influence of Porphyry, see Chapters 2 to 4 in *The Cambridge History of Later Medieval Philosophy*, eds. Norman Kretzmann, Anthony Kenny, Jan Pinborg, ass. ed. Eleanore Stump (Cambridge: Cambridge UP, 1982).

20. As the quotation indicates, Hobbes lacked a sensitivity toward the different kinds of scholasticisms. Aquinas, Scotus, Lombard, Suarez, and others are lumped together. The only concession he seems to make here is that Protestant thinkers like Luther and Melancthon have some sensible opinions. For an example, see Thomas Hobbes, *Questions Concerning Liberty, Necessity, and Chance*, *EW* V, 266–268. His church history in verse consolidates the view. See Thomas Hobbes, *Historia Ecclesiastica*, *OL* V, 349–408. On the latter claim (in the body text), see Schuhmann, "Hobbes and Renaissance Philosophy," in *Hobbes Oggi*, eds. Jean Bernhardt et al (Milan: Franco Angeli 1990), 331–349; Gianni Paganini, "Hobbes, Valla, and Trinity," in *British Journal for the History of Philosophy 11*, no. 2 (2003), 183–218.

21. See Quentin Skinner, *Visions of Politics*, 3 vols. (Cambridge: Cambridge UP, 2002), III, Ch. 2 and Richard Tuck, *Hobbes* (Oxford: Oxford UP, 1989), 9.

22. *Leviathan*, XLVI, 462. Cf. *Concerning Body*, "Epistle dedicatory," viii–ix. Other accounts of this history of spiritual darkness include Thomas Hobbes, *An Historical Narration Concerning Heresy, and the Punishment Thereof*, *EW* IV, 396; Thomas Hobbes, *Writings on Common Law and Hereditary Rights*, ed. Alan Cromartie and Quentin Skinner, *HW* XI (Oxford: Oxford University Press), 93–94; and *Questions Concerning Liberty, Necessity, and Chance*, 63–65. For a more balanced account, see *Six Lessons*, 226 and *Behemoth*, 215.

23. The contrast is present in the term "Aristotelity" Hobbes coins. See *Leviathan* XLVI, 462. Cf. *Latin Leviathan*, 497–499 (473–475). For a brief and perceptive account, see Douglas Jesseph, *Squaring the Circle: The War between Hobbes and Wallis* (Chicago, IL: University of Chicago Press, 1999), 60–61.

24. For a recent study of Hobbes's intellectual milieu in Paris, see Noel Malcolm, "Hobbes and Roberval," in his *Aspects of Hobbes*, 156–199.

25. On Hobbes and Gassendi, see Lisa Sarasohn, "Motion and Morality: Pierre Gassendi, Thomas Hobbes, and the Mechanical World-view," in *Journal of the History of Ideas 46*, no. 3 (1985), 363–379; Noel Malcolm, "Biographical Register

(Gassendi, Pierre)," *HW* VII, 834–836; and G. A. J. Rogers, "Hobbes and His Contemporaries," in *The Cambridge Companion to Hobbes's 'Leviathan,'* 422–424; and Serjeantson, "Hobbes, the Universities, and the History of Philosophy," 126–127.

26. See *Correspondence* I, Letters 62 and 66. Gassendi is listed in the pantheon of true philosophers. See *Concerning Body*, "Dedication," xii. In *Six Lessons* (334) Hobbes writes that a summary of his book ("Ethics demonstrated") was praised by Mersenne and Gassendi. On this work, see Malcolm, *Aspects of Hobbes*, 500–501. Both Mersenne and Gassendi wrote short letters of recommendation that were published in the 1647 edition of *De Cive*. For these, see Thomas Hobbes, *De Cive: The Latin Version*, ed. Howard Warrender (Oxford, Clarendon, 1983), *HW* II, 85–86.

27. See *Six Lessons,* 342. The likely point of disagreement is the Epicurean belief in the existence of void. In addition, in *A Minute of First Draught of the Optiques* (British Library MS Harl. 3360, fols. 101, 106, and 185) Hobbes refers to Gassendi's astronomical treatise, *De apparente magnitudine solis humilis et sublimis* (Paris, 1642).

28. See Noel Malcolm and Jacqueline Stedall, *John Pell (1611–1685) and his Correspondence with Sir Charles Cavendish: The Mental World of an Early Modern Mathematician* (Oxford: Oxford UP, 2005), Part III.

29. Both appear in *The Philosophical Writings of Descartes*, 3 vols., eds. John Cottingham et al (Cambridge: Cambridge UP, 1984), II.

30. The relationship between metaphysics and natural philosophy was a huge issue in early modernity. For a discussion that serves also the specific context at hand, see Steven Shapin and Simon Schaffer, *Leviathan and the Air-Pump: Hobbes, Boyle, and the Experimental Life* (Princeton, NJ: Princeton UP, 1985), Chapter III, which should be read along with the comments by Noel Malcolm, *Aspects of Hobbes*, 187ff.

31. *Leviathan*, XLVI, 463. For other instances of this strategy, see *Concerning Body* I.V.12, 62–63 & I.VI.17, 87; *Behemoth*, 185, 215–217 & 222; and Thomas Hobbes, *Decameron Physiologicum*, *EW* VII, 78.

32. See, respectively, *Concerning Body* II.VIII.23–24, 117–119 and *Decameron Physiologicum*, 72. Cf. *De Motu* XXXVII.3, 404.

33. *Concerning Body* II.VIII.20, 116–117.

34. See, respectively, *Concerning Body*, II.VII.3, 94–94. Cf. *De Motu* XXVIII.2, 332; *Decameron Physiologicum*, 120. Cf. *De Motu* IV.1, 125–126 & XXI.2, 254 and *Six Lessons*, 222.

35. *Leviathan* XLVI, 467. Cf. *De Motu* VI.1–2 & 7, 135–137 & 141 and XXVII.7 & 18, 318–319 & 326; *Concerning Body* I.VI.18, 88–89 and IV.XXX; Thomas Hobbes, *Seven Philosophical Problems*, *EW* VII, Chapter I, esp. p. 7; and *Decameron Physiologicum*, Chapter VIII, esp. p. 143. For Aristotle's account of motion, see Aristotle, *Physica* II and III. The idea criticized here by Hobbes however appears in a clear form in Aristotle, *De Caelo* I.2–3.

36. *Leviathan* XLVI, 463. Cf. *Behemoth*, 215; Thomas Hobbes, *Considerations upon the Reputation, Loyalty, Manners and the Religion of Thomas Hobbes*, *EW* IV, 427; *Answer to Bishop Bramhall*, *EW* IV, 305; and *Decameron Physiologicum*, 81.

37. See *Leviathan* XLVI, 464–465 (where the two quotations in the paragraph also appear). Cf. *Latin Leviathan*, 497–498(473–474). For a discussion, see Paganini, "Hobbes's Critique of the Doctrine," 338–339 & 346–351. For Hobbes's treatment of the errors, falsity, and fallacies that lurk behind the critique, see *Concerning Body* I.V.

38. See, e.g., *Leviathan* XLVI, 466; *Latin Leviathan* XLVI, 499–500(474–476); and *Behemoth*, 217–218. For a discussion, see Shapin and Schaffer, *Leviathan and the Air-Pump*, 92ff.

39. Hobbes's understanding of ethics and moral philosophy is not clear and, almost needless to add, differs from ours. On one hand, ethics and moral philosophy are part of natural philosophy, but on the other hand ethics and moral philosophy has a point of contact with politics because moral philosophy has to do with natural law. The first issue is more pressing here. For a discussion, see Tom Sorell, *Hobbes* (London: Routledge, 1986), 6–7; Richard Tuck, "Hobbes's Moral Philosophy," in *The Cambridge Companion to Hobbes*, ed. Tom Sorell (Cambridge: Cambridge UP, 1996), 175–207.

40. See Thomas Hobbes, *The Elements of Law*, 2nd edition, ed. Ferdinand Tönnies (New York, NY: Barnes & Noble, 1969), I.XIII.3, 66; Thomas Hobbes, *On the Citizen*, ed. and trans. Richard Tuck and Michael Silverthorne (Cambridge: Cambridge UP, 1998) "Preface to the Readers," 7–8; *Concerning Body* I.I.7, 7–10; and *Behemoth*, 217–218.

41. *Leviathan* XVII, 119. Cf. *Elements of Law* I.XIX.5, 103 and *On the Citizen* I.2, 21–22 & 24–25 & V.5, 71. The original argument appears in Aristotle, *Politica* I.2, 1253a1–35, which is of some use to read along with *Historia Animalium* Bk. I, especially 487b32–488a17.

42. This is a tenacious belief. See, for instance, the otherwise laudable discussion by Tom Sorell, "Hobbes's Moral Philosophy," in *The Cambridge Companion to Hobbes's 'Leviathan'*, 146.

43. *On the Citizen* is perhaps firmest in this respect. See *On the Citizen*, "Preface to the Readers," 11.

44. *Elements of Law* I.XVII.1, 87–88. The name Aristotle appears in manuscript copy A of the work (BL, MS Harl. 4235) that Tönnies takes to be the most authoritative. See Ferdinand Tönnies, "The Editor's Preface," in *Elements of Law*, viii–ix. Compare with the metaphor Hobbes presents in *On the Citizen* III.9, 48.

45. Hobbes, *Leviathan* XV, 107. Cf. *Elements of Law* I.XVII.1, 87–88 and *On the Citizen* III.13, 49–50.

46. Aristotle discusses this in *Politica* VI.2 and VII.14. For Hobbes's comments, see *Elements of Law* II.VIII.3, 170; *On the Citizen* X.8, 122; *Leviathan* XXI, 150; and *Behemoth*, 233. For a discussion, see Tuck, "Hobbes and Democracy" and Kinch Hoekstra, "A Lion in the House," esp. 196–197. Both in *Rethinking the Foundations of Modern Political Thought*, eds. Annabel Brett, James Tully, and Holly Hamilton-Bleakley (Cambridge, Cambridge UP, 2006).

47. See *Elements of Law* II.X.8, 189; *On the Citizen* XII.3, 133–134; *Leviathan* XLVI, 471; and *Behemoth* 218.

48. See, e.g., *Leviathan* XLVI, 469.

49. Aristotle, *Politica* I.2, 1253a7–18.

50. Aristotle, *Ethica Nicomachea* I.13, 1102a5–17.

51. *Leviathan* XI, 70. Cf. *Leviathan* VI, 46 and *Elements of Law* I.VII.6–7, 29–30 & I.IX.18, 47. Note that Hobbes's critique is not uncommon. In Tudor Oxford "objection to Aristotle is sharpest on the doctrinal issue of the good life [...] and it rehearsed familiar arguments of Christian controversy against the pagan notion of the *summum bonum*." See McConica, "Humanism and Aristotle in Tudor Oxford," 308. McConica is referring to Rainolds's *An excellent oration of that late famously learned Iohn Rainolds, D.D. and lecturer of the Greek tongue in Oxford Very usefull for all such as affect the studies of logick and philosophie, and admire profane learning*, tr. by John Leycester (London, 1638) which was given in 1573 and the original Latin text first printed in *Orationes duodecim* (London, 1619). The critique should also be read in the light of Hobbes's rebuttal of the Aristotelian doctrine of four causes and, in particular, of the final cause. See Hobbes, *Concerning Body* I.III.20, 43–44 & II.X.7, 132. Note also that the denial of teleology in physics and ethics is not as remote as it may appear, for both are part of natural philosophy.

52. *Questions Concerning Liberty, Necessity, and Chance*, 194.

53. *Elements of Law* I.IX.21, 47.

54. This finds some support from the comment Hobbes makes in *Decameron Physiologicum*, 83.

55. *Elements of Law*, I.XIV.13–14, 74. Cf. *On the Citizen* I.7&15, 27&31 & V.4, 70–71 and *Leviathan* XIV, 91–92.

56. For a valuable discussion, see Sorell, "Hobbes's Moral Philosophy."

57. Here I own a great deal to Donald Rutherford's "In Pursuit of Happiness: Hobbes's New Science of Ethics," in *Philosophical Perspectives 30*, no. 1(2003), 369–393.

58. See *Leviathan* XIII, 90. Cf. *Elements of Law* II.IX.4, 180 and *On the Citizen* XIII.15, 151.

59. *Leviathan* XXI, 152. Cf. *Elements of Law* II.IX.4, 180–181 and *On the Citizen* XIII.15, 150–151. I need to add that my conclusion has a touch of optimism and leaves out the important question how an individual can protect him- or herself from the arbitrary use of power by a sovereign.

Chapter 12

Hobbes and Aristotle on Biology, Reason, and Reproduction[1]

Ingrid Makus

Hobbes is famously critical of Aristotle's politics, claiming that it is grounded in philosophical, metaphysical, and scientific error.[2] In particular, Hobbes challenges Aristotle's idea that the "human good" is embedded in nature and that political association is natural to human beings.[3] On Hobbes's account, all human activity, including the construction of human goods can be explained best through his own "corrected" mechanistic-materialist, individualistic "science" of human behavior. Politics, Hobbes wants to demonstrate, is the result of artificial covenant rather than of the natural outgrowth of political community.

Hobbes's own attempt to found a new "science" of politics that challenges Aristotelian assumptions is flawed, however. In particular it cannot adequately explain a most basic and minimal human good—the perpetuation of the human species and the reproduction of the political community. In Part One of this paper I provide a brief overview of the limitations of the Hobbesian model. Such an overview is useful in highlighting what is similarly deficient or contradictory in Aristotle's efforts to ground politics in a "science" of biology, explored in Part Two of the paper. The limitations of Hobbes's "political science" center on his depiction of familial relations which he aims to establish on the basis of artificial covenant. Seemingly, Aristotle's emphasis on the natural biological basis of family relations would exempt him from such difficulties. Yet, there are similar difficulties—both attempt to justify infanticide for example. Recognizing the limitations of the Hobbesian and Aristotelian models can lead to a fuller discussion of contemporary issues surrounding the politics of biology and reproduction.

PART ONE

Hobbes on Reason and Science

Hobbes' depicts human reason as an activity that entails a *succession* of thoughts (mental discourse).[4] This activity can be relatively unguided, in which case it resembles dreaming. When it is guided by desires or appetites, however, it becomes more sequential. Both humans and animals engage in the kind of regulated train of thought that seeks to find the means to achieve a desired effect. But only human beings have the curiosity to seek to discover the causes and consequences of desired effects and to find diverse means to achieve desired ends.

Another distinctive human activity is Speech which consists of using agreed definitions, or names, to signify things, and to connect them in a causal sequence. Those things that are capable of being named are the only things that can be taken as the subject of reason. These include (1) what is deemed Matter or Body (2) names by which matter and body are distinguished from one another (3) things that bring into account properties of our own bodies, which are derived from what affects our five senses. (4) Names of the names themselves.[5] The operation of human reason is the same, whether the tools or objects of reckoning are numbers, words, or "political entities"—which are themselves only representations of words. Reasoning becomes calculation in Hobbes—"in what matter soever there is place for *addition* and *subtraction,* there is also place for *Reason;* and where these have no place, there *Reason* has nothing at all to do."[6]

Science is both the activity of reasoning and the end result of reasoning. Hobbes introduces it as a method that seeks the "truth" by using agreed upon definitions of words, or speech,[7] and putting these definitions together in a causal order. It is a way of calculating the relationship among things with the aim of causing and predicting desired consequences—"And whereas Sense and Memory are but knowledge of Fact, which is a thing past, and irrevocable; *Science* is the knowledge of Consequences, and dependence of one fact upon another."[8] The "truth" that can be acquired as the result of Science is not absolute but conditional in the sense that it rests on making the proper logical connections among agreed upon definitions, definitions which can change. Hobbes declares that "True and False are attributes of Speech, not of Things. And where Speech is not, there is neither Truth nor *Falsehood.*"[9] For Hobbes there is no outside reference point or Aristotelian "essence" in Nature to which speech names refer. They are simply agreed upon definitions.

Hobbes on Biology and Politics

What moves human beings to activate their reason, or to engage in any activity, are their appetites and aversions, grounded in the biological body. Both humans and animals, Hobbes maintains, are propelled by two sorts of motion (1) the *vital* or involuntary motions involved in bodily activity such as breathing and (2) the *voluntary motions* that propel toward something (appetites) or away from something (aversions).[10] What distinguishes humans from animals is the ability to name their appetites and aversions. Most important for Hobbes's account of "politics" is that individual human beings can and do name what they desire as "good" and what they want to avoid as "evil." In the absence of an agreed-upon definition of names and words, it cannot be decided by reference to "Nature" or a transcendental "essence" what is "true" and "false," "good" and "evil." These require an Arbiter or Judge who determines among conflicting definitions what is to be considered "true" and "good" through establishing laws backed by punishment for breaches. Only a Sovereign Will with absolute authority has the power to establish and compel obedience to such laws, a situation which by Hobbes's definition constitutes "civil society." Similarly, the enterprise of Philosophy requires a common Arbiter or Judge, a characteristic of civil society. The outcome of Science is wisdom or philosophy—"By PHILOSOPHY, is understood the *Knowledge acquired by Reasoning, from the Manner off the Generation of any thing, to the Properties; or from the Properties, to some possible Way of Generation of the same; to the end to bee able to produce, as far as matter and humane force permit, such Effects, as humane life required.*"[11] Hobbes's own philosophy of politics, however, cannot provide a plausible or reasonable account of how human beings reproduce themselves and ensure generational continuity.

Hobbes consistently applies his own reckoning of what constitutes human reasoning and the activity of science to politics. All human behaviors and relations, whether political, familial, or social, are defined as driven by similar biologically grounded appetitive (and aversive) imperatives. Hobbes uses the device of the "state of nature" to demonstrate how individual human beings, without a Sovereign Will who can establish and compel obedience to civil law, are necessarily driven by their appetites and aversions to a condition of war of all against all. This is a state where there is no industry or science. There can be no shared knowledge of "truth" or "good," since there is no common authority to determine what signifies an agreed upon definition of a thing. Indeed, if one works through Hobbes's logic on the biology of reproduction, there is no guarantee that the human species can reproduce itself. It becomes a state where not only the life of the individual human being, but also the life of the human species is precarious.

The Hobbesian State of Nature

According to Hobbes's own abstract definition of the state of nature as a hypothetical state that can occur at any time, it is a condition in which is there is no common Sovereign authority to establish and compel obedience to civil law.[12] In this condition, sexual appetite may drive male and female to unite but it is unlikely that the male would remain with the female afterward, or even acknowledge "paternity" of common offspring. The conjugal relationship is a result not of "nature" or "biology," but of civil law, Hobbes proposes. Only in civil society, where a Sovereign Authority erects Marriage Laws, can one assume continuity in male-female associations.[13] In the state of nature, relations between male and female, and, more importantly, between adults and offspring, are subject to similar imperatives as relations between strangers—competition, self-interest, and a potential war of all against all. "Generation" or biological ties do not seem to guarantee or engender any unique sort of bond. This emerges most clearly in Hobbes's presentation of the dynamic between and a woman and her offspring.

In the "state of nature," the consummation of male and female sexual desire may lead to pregnancy and birth. But, there is little justification, based on reason and science, for why the woman who gives birth would feed and care for the infant, thereby giving it life. Indeed, Hobbes asks whether she might not abandon it and leave it to another to find and nourish:

> For in the condition of meer Nature, where there are no Matriomoniall lawes, it cannot be known who is the Father, unlesse it be declared by the Mother: and therefore the right of Dominion over the Child dependeth on her will, and is consequently hers. Again, seeing the Infant is first in the power of the Mother, so as she may either nourish, or expose it, if she nourish it, it oweth its life to the Mother; and is therefore obliged to obey her, rather than any other; and by consequence the Dominion over it is hers. But if she expose it, and another find, and nourish it, the Dominion is in him that nourish it.[14]

Using Hobbes's own reasoning, based on his depiction of human behavior in the state of nature, it is likely that the mother would abandon rather than nourish the infant. In the absence of an effective Law of Gratitude which compels individuals not to harm those who have conferred benefits on them, there is no guarantee that the child would not become the mother's enemy. There is even less incentive for a stranger to nourish a child in the absence of such a guarantee. Since human infants are born dependent requiring the care of another to live, and one cannot assume that care is forthcoming in the state of nature, even the propagation of the human species is precarious.

What's Missing in Hobbes?

Several elements are missing in Hobbes's biological scientific explanation for human behavior. (1) He briefly mentions a "natural inclination" of parents to their offspring.[15] However this seems to manifest itself in the desire to increase one's own honor and power through one's kin[16] rather than a bond that ensures proper care for one's offspring. There is no comprehensive account of a maternal or paternal instinct or desire to have and care for children or a moral relationship between them that exists outside of contractual exchange, that is a reckoning based on voluntarily pursuing one's own interests or appetites. More serious than the lack of paternal instinct is the lack of a maternal bond that operates outside of contractual considerations.

(2) Hobbes makes few distinctions between male and female in their association with their offspring. Biological imperatives of conception, pregnancy and birth give the "mother" knowledge that the infant is her "own." They also give her the first opportunity to have access to the new life—to exercise power that comes with the potential to nourish the infant, thus establishing dominion over it. However, absent in Hobbes's description of male and female biology or behavior is any notion of a "natural" maternal instinct or love of one's own offspring. In fact, Hobbes presents women as having similar capacities for reasoning as men. Both sexes can calculate to kill another in the state of nature, if they so desire.[17] And, given the dynamics of the state of nature, it is likely that the woman's reckoning would lead her to abandon rather than nourish the infant. The distinction between abandonment and infanticide has little meaning in the state of nature. The biological fact that humans are born requiring nourishment from another to live means that the failure to provide such nourishment is akin to causing death. Furthermore, infanticide carries no moral condemnation in a state of nature, since what is "good" and "evil" are meaningless in the absence of agreed upon definitions.

(3) There is little concept of the uniqueness of biological ties or love of one's own in Hobbes "science of politics." Instead, familial relations, including relations between parents and children, are assumed to be as artificial as political relationships. The parent-child relationship, as Hobbes depicts it, is based on an artificial covenant, similar to the political covenant that generates a Sovereign Will in civil society. A covenant requires that one party perform "afterwards" their part of the contractual exchange of benefit. Yet covenants are only active where there is a Sovereign Will to ensure obedience, found only in civil society. Generational continuity remains a problem in civil society to the extent that the family—as a "private association" is left unregulated by the Sovereign Will. As such the relations between parents and children are left to the exigencies of the state of nature—a competitive potential war of all against all.

PART TWO

Aristotle on Biology, Generation, and Politics

Seemingly, Aristotle overcomes the limitations of the Hobbesian model in accounting for the perpetuation of the species and the reproduction of the political community by fleshing out precisely what Hobbes attempted to delete:

(1) Aristotle grounds political community on a notion of male and female coming together in a natural association that unites their differences and propagates both the species and the political community.[18] In the opening chapters of *The Politics* Aristotle explains the beginnings of political community by employing a method he deems to be capable of leading to scientific knowledge—a method that considers things as developing or growing out of their most basic components. The most basic elements of political association are the male and female. They unite according to a natural compulsion, not merely to have intercourse, but also to procreate. The impulse in humans, as in animals, is a biological imperative to leave behind something of themselves (in contemporary terms one could say their "genetic material"). According to Aristotle—"First of all, there must necessarily be a union or pairing of those who cannot exist without one another. Male and female must unite for the reproduction of the species—not from deliberate intention, but from the natural impulse, which exists in animals generally as it also exists in plants, to leave behind them something of the same nature as themselves."[19] Male and female unite to procreate and form the household (along with slaves). The association of households leads to villages that combine to form the larger political community. Aristotle presents this progression as "natural" in the sense that the move toward political community is the final end of the development of human nature.

According to Aristotle's notion of *physis,* all living things, including human beings, are engaged in a process of "becoming" or movement toward the fulfillment of their naturally given potential or essence. The "nature" of human beings, Aristotle insists, can only reach its fulfillment in a form of political association. Some animals share with humans a "gregarious" nature, and most share the ability to make "sounds" that "indicate pleasure and pain." But only humans have the "faculty of language." The defining quality of language and of what distinguishes humans from other animals is that language communicates what is just and unjust, good and evil—"It is the peculiarity of man, in comparison with the rest of the animal world, that he alone possesses a perception of good and evil, of the just and the unjust, and of other similar qualities; and it is association in [a common perception of] these things which makes a family and a polis."[20] In agreement with Hobbes, language also declares what is advantageous. In contradiction to Hobbes, what is advantageous by Aristotle's definition is analogous to what is just.

Aristotle also deems the polis to be "prior" to the family and the individual in the sense that the capacities and functions of the individual and the family, or the character of the parts, cannot be developed without the existence of the "polis" or the presence of the whole; nor can the individual (the part) exist without the polis (the whole). Most importantly, in contrast to Hobbes, who insists that what a thing *is* is simply the name that is given to it (by agreed upon definition), Aristotle maintains that a thing may have a name that is not analogous to what it *is*—"All things derive their essential character from their function and their capacity; and it follows that if they are no longer fit to discharge their function, we ought not to say that they are still the same things, but only, by an ambiguity, they still have the same names."[21] The purpose of language for Aristotle is to further "moral prudence and virtue," although it is sometimes used for injustice. For Hobbes, humans are likely to use language to deceive others about what is "just" and "unjust." For Aristotle, humans have a natural impulse toward political association and the will to construct a political association that propels them to perfect their nature. Hobbes takes Aristotle to task for precisely this, proposing that humans have a natural impulse toward competition for honor which propels them to war. The notion that living beings have an essential nature is central to Aristotle's distinction between male and female.

(2) Unlike Hobbes, Aristotle distinguishes between male and female as biological entities with distinctive relations to each other and to their offspring.[22] In *On The Generation of Animals* Aristotle prefaces his discussion of how species reproduce with a clear statement on his teleological essentialism—"There are four causes underlying everything: first, the final cause, that for the sake of which a thing exists; secondly, the formal cause, the definition of its essence (and these two we may regard pretty much as one and the same); thirdly, the material: and fourthly, the moving principle or efficient cause."[23] He is also clear that that the distinction between the "male" and "female" principles in their essence is that although both are necessary to generation, the male contains the "efficient cause" or the moving principle of generation and the female contains the material or the raw matter of generation. The "scientific proof" Aristotle provides of his supposition has to do with a lengthy discussion of the role that semen or "seed" plays in procreation. He questions and challenges the theory, which he attributes to some of the ancient natural philosophers, that the semen or the seed that creates the new life is generated by both the male and the female, and that the semen or seed is secreted by the whole body. He rejects both suppositions, drawing on his observations of male and female animal and human coition and procreation. His discussion

centers on "four proofs" relating to (1) the intensity of the pleasure of coition, (2) the inheritance of genetic mutations, (3) the resemblance of offspring to the parents, and (4) the relationship of the parts to the whole.

Aristotle rejects the argument that the female emits any form of semen or that the female "catamenia" or menstrual blood is a form of semen, on the grounds that the female can conceive without the sensation of pleasure in intercourse. The female contributes through the catamenia the material for generation, whereas the male contributes the principle of movement through the male ejaculate. (Female ejaculate does not have the generative properties of male ejaculate, Aristotle maintains). The female secretion in the form of menstrual discharge is not fully formed enough to provide the main principle of generation. Aristotle compares a woman to an "impotent male." It is her incapacity that makes her female. Furthermore, the male stands for the active, the female for the passive element.

In the opening of Book 2 of *On The Generation of Animals,* Aristotle is also clear that the male principle which provides the form and moving principle of generation is more divine and noble than the female principle which provides the matter—"For the first principle of the movement, or efficient cause, whereby that which comes into being is male, is better and more divine than the material whereby it is female. The male, however, comes together and mingles with the female for the work of generation, because this is common to both."[24] Female incapacity is related to "coldness," male capacity to "hotness." For procreation to occur there must be a proper balance of hot and cold. Fertility rates may vary depending on environmental conditions such as climate, diet, and drinking water which can affect the balance of hot and cold in the male-female union. Aristotle presents the female as a necessary but deficient type. In a passage where he notes that males are more often born "defective" than females, he also observes that females mature more quickly after birth than males. He attributes this to their inferiority of form: "For females are weaker and colder in nature, and we must look upon the female character as being a sort of natural deficiency. . . . for all inferior things come sooner to their perfection or end, and as this is true of works of art so it is of what is formed by Nature."[25]

(3)Another distinction between male and female is that the mother is more prone to affection for the offspring than the father. Unlike Hobbes, who portrays relations among family members as similar to relations among strangers, Aristotle emphasizes the uniqueness of the familial bond. He describes a form of friendship between man and wife as having its existence in nature, not only for the purposes of reproduction, as in animals, but also for "various purposes of life." In the household where husband and wife have divided functions "they help each other by throwing their peculiar gifts into the common stock."[26] This friendship has both utility and pleasure and, if both parties are good, it also has virtue. Children become a common bond that unites them further—"For parents love their children as being a part of themselves, and children their parents as having themselves originated from them."[27] The natural affection between parents and children forms a bond that is even stronger in the mother than the father.[28] In Aristotle, unlike Hobbes, knowledge of one's own and the biological connection between parents and offspring generate a moral relationship that obliges parents to care for their children's well-being, thereby, seemingly providing a natural foundation for the continuity and reproduction of the political community. Yet there are serious deficiencies and contradictions in the Aristotelian model, particularly in the way he grounds his politics of reproduction in natural biology and science.

Deficiencies in the Aristotelian Model

(1)Aristotle's presentation of the female as biologically deficient contradicts or undercuts his presentation of maternal love. Human beings love more what they have won by labor, and what they acknowledge as their own. Aristotle acknowledges that the mother labors more than the father to bring forth the child and has more certainty than the father that the child is indeed hers—"These are the reasons, too, why mothers are fonder of their children than fathers; bringing them into the world costs them more pains, and they know better that the children are their own."[29]

In Book 8 of the *Ethics*, Aristotle lists friendship between husband and wife, parents and children as resting on distinctive virtues and functions. Parent-child relations are relations of inequality requiring that the love be proportionate to the goodness of the loved one. In the case of the love of father over children, for example, the father is owed a greater benefit as being the superior one. But Aristotle singles out, over and above paternal love, the love of the mother for the child as an example of the essence of friendship as a good in itself. This rests precisely on the virtue of *actively* loving another rather than *passively* being loved—"Further, love is like activity, being loved like passivity; and loving and its concomitants are attributes of those who are more active."[30] Aristotle describes the "delight mothers take in loving; for some mothers hand over their children to be brought up, and so long as they know their fate they love them and do not seek to be loved in return (if they cannot have both), but seem to be satisfied if they see them prospering; and they themselves love their children even if owing to their ignorance give them nothing of a mother's due."[31] The depiction of women as mothers exemplifying the virtues of maternal love requires that they be *active* in loving their children, rather than *passive* in being loved by them, a quality which is incompatible with their essence as described in Aristotle's biological scientific account of male/female differences in reproduction.

(2) The limitations of Aristotle's model also emerge when one considers his recommendations on infant abandonment. In Book 2 of *The Politics* Aristotle provides a well-known criticism of Plato's community of wives and children as laid out in Book 5 of *The Republic.*[32] Strikingly, Aristotle fails to take Plato to task for the proposal that infants who are "deformed" may be exposed at birth. Aristotle proposes that laws stipulating that "deformed" infants be exposed are necessary for the well-being of the community.[33] As well, he recommends making laws that support inducing "miscarriage" in early pregnancy for the purpose of regulating population growth. However, these recommendations contradict and challenge his claims about parental affection, in particular maternal affection. If women are naturally inclined to love their offspring and care for them, more so than men, how does one explain how they would deliberately and willingly abandon their own offspring to death (rather than to the care of another which might be justified as being beneficial to the well-being of the child) or induce abortion of their own potential offspring? If the father, as the head of the family who rules over the wife in Aristotle's account of the household, simply commands the removal of the "deformed" infant after birth, or the state simply regulates that miscarriage be induced without female agreement, does this not go against female "nature" and thus "Nature" as a whole?

According to Aristotle's "science" of nature, all things contain a potential end or "essence" toward which they move. All substances, including human beings, are in a state of becoming—moving from their potential to their final end, which forms the natural completion of their being.[34] "Deformed" infants might presumably be considered to be "against nature" or nature's mistakes if they are deemed incapable of fulfilling their "human" potential. Yet, this justification for infant exposure is incompatible with Aristotle's emphasis on the soul rather than the body as a "true" indicator of one's nature.[35] It raises the question of how human beings can distinguish what is "against Nature" and what is compatible with the natural development of the essence of a living thing. Aristotle's discussion of "deformed offspring" does not address the problem of how one distinguishes the "nature" of a thing before it is fully formed.

In Book 4 of *On The Generation of Animals,* Aristotle observes that some children resemble their parents, some resemble other relations, some resemble neither parents nor other relations but are still human, whereas others are not even human. In fact, Aristotle proposes that these "monstrosities" of nature, in being a deviation from the ideal are not that far removed from other kinds of "deviations," such as those children who do not resemble their parents and those fetuses that remain female. The "defective female" is on a spectrum with the "defective" offspring—"For even he who does not resemble his parents is, already in a certain sense a monstrosity; for in these cases Nature has in a way departed from the type. The first departure is indeed that the offspring should become female instead of male; this, however, is a natural necessity."[36] Females are a deviation in the sense that they represent the male principle not prevailing over the female principle. This is necessary for procreation of the species. However, the deviation that is a "monstrosity" is not part of the final cause or end of procreations—it is accidental. Yet, by Aristotle's own account of the distinction between what is a random accident, and what is a genetic deviation that might be "useful" to a final end of reproduction of the species, there is no clear way of being able to use scientific observation to tell the differences between the two. In a lengthy passage, Aristotle attempts to explain these distinctions:

> For the monstrosity belongs to the class of things contrary to Nature, not any and every kind of Nature, but Nature in her usual operations; nothing can happen contrary to Nature considered as eternal and necessary, but we speak of things being contrary to her in those cases where things generally happened in a certain way but may also happen in another way. In fact, even in the case of monstrosities, wherever things occur contrary indeed to the established order but still always in a certain way and not at random, the result seems to be less of a monstrosity because even that which is contrary to Nature is in a certain sense according to nature, whenever, that is, the formal nature has not mastered the material nature. Therefore they do not call such things monstrosities any more than in the other cases where a phenomenon occurs habitually.[37]

Only in cases in which the physical deformity is such that organs of life are threatened is it clear that "Nature" intended the death of the infant. To abandon infants (mostly males, since Aristotle observes that males are more often born "defective" than females) without knowing that they are indeed "accidental monstrosities" would go against Nature.

What is eternal and necessary in Nature is a form of knowledge than cannot be ascertained by scientific observations and proofs, according to Aristotle. Accepting Aristotle's account of the essence of male and female then cannot be disclaimed on the grounds of empirical observation of male and female reproductive behavior. Yet it cannot explain what Aristotle presents as a natural outgrowth in social and political behavior which takes the form of a moral relationship among family members which is necessary to the development of the political community.

The human good for Aristotle can only be fully developed through the political community which is deemed to be the "natural end" of human relations. For Aristotle, human reason is deliberative—it aims at the good, including the good of the political community. One might justify the exposure of "deformed" individual infants on the grounds that it aims toward the greater end of the development of the political community, by generating a healthy citizenry. However, one must then deem "nature" to have given women the deliberative capacity to "rule" their instinct of maternal love for the good of the political community. Yet Aristotle is adamant that women's deliberative capacity is not fully developed.[38] The biological distinctions he makes between male and female present a portrait of women as being "incomplete" in this capacity, and as being "without authority" in the context of *The Politics.*

In both Aristotle and Hobbes, the mother-child association is of greater import for generational continuity than the father-child one. Aristotle shows us what is missing in Hobbes—a notion of maternal love and care for one's own. Hobbes shows us what is missing in Aristotle—an acknowledgement of the deliberative or reasoning capacity of women. Both use the claims of "science" in the form of biological imperatives to depict relations between mothers, fathers and children. However, neither is successful in demonstrating how these relations lead to a politics where, in Hobbes's case, artificial covenants ensure biological and political perpetuation of the species, or where, in Aristotle's case, natural human bonds ensure that the good of the political community is furthered. Each requires elements of the other to provide a more complete account of a politics that can supply the fundamental human good of political continuity. Consideration of these models is vital to reflection on contemporary debates where the Hobbesian problem of considering reproduction as subject to rational self-interested considerations exists along side the Aristotelian contradiction of presenting women as having a greater maternal love for offspring while maintaining that the science of biology aims for genetic manipulations for the purpose of achieving the "perfect natural" form of the human.

ENDNOTES

1. An earlier version of this paper entitled Aristotle and Hobbes on Science as the Basis of the Human Good was presented at the 11th International Conference of the ISSEI in Helsinki, Finland, July 2008.
2. Thomas Hobbes, *Leviathan* (Middlesex, England: Penguin, 1977), 688–89.
3. Hobbes, *Leviathan*, 225–26.
4. Hobbes, *Leviathan*, 94.
5. Hobbes, *Leviathan*, 107.

6. Hobbes, *Leviathan*, 111.
7. "So that in the right Definition of Names, lyes the first use of Speech; which is the Acquisition of Science:" Hobbes, *Leviathan*, 106.
8. Hobbes, *Leviathan*, 115.
9. Hobbes, *Leviathan*, 105.
10. Hobbes, *Leviathan*, 178,
11. Hobbes, *Leviathan*, 682.
12. Hobbes, *Leviathan*, 185-86.
13. Hobbes, *Leviathan*, 253.
14. Hobbes, *Leviathan*, 254.
15. Hobbes, *Leviathan*, 253.
16. Hobbes, *Leviathan*, 254.
17. Hobbes, *Leviathan*, 253.
18. Aristotle, *Politics*, translated Ernest Barker (London: Oxford University Press, 1981), 2–3.
19. Aristotle, *Politics*, 3.
20. Aristotle, *Politics*, 6.
21. Aristotle, *Politics*, 6.
22. Aristotle, *On the Generation of Animals*, translated by Arthur Platt (eBooks@ Adelaide, 2007), book 1, 1–2.
23. Aristotle, *On Generation*, book 1, 1.
24. Aristotle, *On Generation*, book 2, 1.
25. Aristotle, *On Generation*, book 4, 23.
26. Aristotle, *Nicomachean Ethics*, translated by David Ross (Oxford: Oxford University Press, 1980) 214.
27. Aristotle, *Ethics*, 212–13.
28. Aristotle, *Ethics*, 213.
29. Aristotle, *Ethics*, 234
30. Aristotle, *Ethics*, 234.
31. Aristotle, *Ethics*, 205.
32. Aristotle, *Politics*, 43–47.
33. Aristotle, *Politics*, 327.
34. "We must not say that each is of a certain quality because it becomes so, but rather that they become so and so because they are so and so, for the process of Becoming or development attends upon Being and is for the sake of being, not vice versa." Aristotle, *On Generation*, book 4, 1.
35. See Aristotle's discussion of the differences between "natural" and "conventional" slavery in the *Politics*, 13–14.
36. Aristotle, *On Generation*, book 4, 8.
37. Aristotle, *On Generation*, 14–15.
38. Aristotle, *Politics*, 35.

Part IV

THE SCIENTIFIC ROOTS OF LIBERALISM AND CONTEMPORARY "BIOPOLITICS"

Chapter 13

Locke and the Problematic Relation between Natural Science and Moral Philosophy

Lee Ward

A recent exchange between Jeremy Waldron and Michael Zuckert highlighted an important, but often neglected, aspect of John Locke's political philosophy having to do with his conception of the relation between natural science and moral philosophy. In his book *God, Locke, and Equality,* Waldron examined the problematic status of the idea of species in Locke's thought.[1] On the one hand, in Locke's political writings the doctrine of natural equality appears to rely on the fact that human beings are all "Creatures of the same species."[2] Yet in his most philosophic work, the *Essay Concerning Human Understanding,* Locke systematically dismantled the concept of natural species and reduced the notion of species to arbitrary linguistic categories with no real connection to the essence of natural substances. Waldron insists that in order to save Locke's doctrine of equality from the incoherence produced by his epistemological skepticism, we need to recognize the indispensable theological foundation of Locke's idea of equality, an idea ultimately rooted in the human being's intellectual capacity to conceive of God and the moral law.

In his critique of Waldron, Zuckert argues that Waldron exaggerates Locke's skepticism about the idea of species.[3] Zuckert also maintains that the religious basis of Locke's doctrine of equality is more complex than Waldron recognizes. Zuckert's Locke grounds equality in two distinct kinds of arguments: both a rationalist natural theology and a pietistic belief in equality based on all humanity being the "Workmanship" of the same creator God. The heart of what Zuckert calls "Locke's wager" is the proposition that, in lieu of demonstrative philosophic proof of the existence of God and immortality of soul, these two kinds of arguments with distinct sorts of evidence would together provide a compelling case for equality to the broadest range of readers.

This chapter proposes to examine Locke's account of the relation between natural science and moral philosophy by following the tracks laid by Waldron and Zuckert, as well as cutting new trails of our own. Our approach is to partly, and only temporarily, set aside the theological questions that interested Waldron and Zuckert, and to focus rather initially on Locke's treatment of natural science and its problematic relation to moral philosophy. Only once this relation is clarified will we return to the ethereal realm, hopefully well equipped to locate properly the relation of natural and moral philosophy in the context of Locke's political philosophy. We will suggest that both Waldron and Zuckert are correct, albeit in different ways. Waldron is right to see the problematic effects of Locke's natural philosophy as it relates to his moral philosophy. However, Zuckert is correct to observe that Locke's moral teaching presupposes a conceptual flexibility embracing two forms of knowledge with distinct sorts of evidence. The Lockean "wager" we will explore, however, has not to do with complementary forms of natural and revealed theology, but rather with the uneasy, even tenuous, association between demonstrative moral philosophy and experimental natural science that binds the diverse intellectual elements of Lockean society. This study will argue that there is a divided mind built into Locke's thought—a tension between his conceptions of moral and natural philosophy that holds potentially serious political and philosophical implications for liberal society.

THE PROBLEM OF SPECIES

As is well known, one of the bedrock principles of Locke's epistemology is that *contra* scholastic and Cartesian philosophy, there are no innate ideas. Not only does the human mind lack an innate idea of species membership, it is originally a "White Paper" or "empty cabinet" devoid of any ideas whatsoever.[4] The only source of ideas is experience in the form of sensation produced by external sensible objects or reflection upon the internal operation of one's own mind (2.1.1.104). The three kinds of ideas produced by sensation and reflection are simple ideas formed by "qualities that affect our senses" (2.2.1.119), complex ideas composed of multiple simple ideas, and a third kind Locke terms mixed modes. Mixed modes are the most abstract form of complex idea because they exist at the furthest remove from nature. Whereas nonmodal ideas, simple or complex, depend on a natural archetype, modes are solely products of the human mind's capacity to discern relations and separate ideas by abstraction. Modes are free from dependence on nature, and are thus not to be "looked upon to be characteristical marks of any real Beings that have a steady existence" (2.22.1.288). Geometric forms such as a triangle, or moral terms such as gratitude and homicide do not, Locke claims, depend on the existence of anything in nature conforming to or even resembling these ideas (2.12.4.165). Freed from the heuristic properties of natural archetypes, mixed modes are purely of human invention. However, the connection between mixed modes and simple ideas, while attenuated, is still foundational. Modes are constructed by mind on the basis of material supplied by distinct simple ideas (2.22.3.289). More importantly, Locke relates that mixed modes are the kinds of complex ideas that "are the greatest part of the words made use of in Divinity, Ethics, Law, Politics and several other Sciences" (2.22.12.294). For Locke, ideas not only explain how we can know anything, but also what we can know and the degrees of certainty attainable.

The primary distinction between what can and cannot be known emerges in Locke's discussion of substance. Here Locke claims we can have no clear idea of substances. Far from being the foundational principle of intelligibility as it is for Aristotle and the classical philosophical tradition, Locke demotes the meaning of substance to being the vague term people use to describe the unknown thing that holds corporeal bodies together (2.23.2, 4.296-7).[5] Our ignorance of the real nature of substance is due to the human mind having little idea what unites atoms into form and practically no sense of the infinite variety of potential relations and powers in any, even apparently the most simple, substance (2.23.25-6.309-10). This presumption of ignorance partly derives from Locke's commitment to a corpuscular account of nature according to which we have "but very obscure and confused ideas of . . . minute bodies (2.29.16.370)." The "perfectest" idea of a particular substance of which we are capable is the product of probability and judgment based on collecting "most of those simple ideas, which do exist in it (2.23.7.299)."[6] However, Locke insists that there is no way to know all the possible powers possessed by a substance.

The implications of the ineluctable essence of substance for Locke's philosophy are enormous. The boundaries of human knowledge are, on the one hand, the substratum of physical nature in which "there is not a Plant or Animal that does not confound the most inlarged understanding," and, on the other hand, the immaterial superstratum of angels and such in which for all we know there may be "more Species of Intelligent Creatures above us" than material beings below (3.6.9.444; 3.6.12.447). More importantly, Locke maintains that our intrinsic ignorance of substance means that human understanding is constitutionally incapable of proving or disproving the immortality of soul. Of this crucial theological, and by extension political, doctrine, Locke concludes "it becomes the Modesty of Philosophy, not to pronounce magisterially, where we want that evidence that can produce knowledge" (4.3.6.541-2). The corpuscularian hypothesis can neither completely explain mental activity internal to mind, nor can it support or refute metaphysical speculation about soul. Locke is thus fully aware of the limits placed on natural philosophy by the perceiver-dependent basis of our understanding of substance.

In Locke's radically particularistic ontology "All things that exist" are "particulars" (3.3.1.409), and as such the proper philosophic stance toward being requires resisting the temptation to deduce general sorts or species from presumed essential qualities.[7] The key is to recognize the problem of language whereby one word can be used to mark a multitude of particular existences. According to Locke, it is the inherent arbitrariness of words as human constructs that compels any truly philosophic perspective to involve distinguishing the "nominal essence"—the name we give to a sort of thing displaying certain observable characteristics—and the "real essence" or its internal constitution, which is, given the limited cognitive capacities of human beings, as irreducibly mysterious as the internal machinery of the Clocktower at Strasbourg to the uneducated peasant in the field (3.6.3.440). Of course, the impossibility of achieving perfect identity between the nominal and real essence of substance contrasts, at least in the formal sense, with the situation relating to mixed modes in which they are in principle identical inasmuch as modes are themselves entirely constructs of mind.

Whereas mixed modes can in principle be known with certainty precisely because they are constructs of mind indistinguishable from their linguistic representation, in natural science language can facilitate improved understanding only when it is acknowledged to offer merely descriptions of simple ideas, rather than real essences of things, and thus can never provide more than hypothetical accounts of observable natural phenomena. Locke's argument thus presumes the provisional, and at best probabilistic, character of natural science.

What does Locke's account of substance mean for a moral philosophy ostensibly grounded on membership in a common species? Even if we set aside for a moment the theological concerns of Waldron and Zuckert, Locke's idea of substance remains problematic in a number of senses. First, Locke clearly rejects the Aristotelian conception of species and real essences as key logical categories for defining natural sorts or kinds of things. In Aristotelian moral philosophy the human being is a certain kind of substance with a distinct ontological status: we can speak with certainty about the moral properties of human beings because we are part of an intelligible natural order composed of distinct species of things. For Locke, however, species are primarily a matter of semantic value inasmuch as while nature may admit of observable resemblances among things, the names of species are purely conventional due to the arbitrariness of words. Resemblances may provide a minimal natural standard of intelligibility, but as Ayers observes resemblances "do not draw lines" or establish boundaries.[8] Locke warns against simply eliding phenomenally similar individuals into a structurally defined class for "we shall find everywhere, that the several Species are linked together, and differ but in almost insensible degrees" (3.6.12.447).

As Locke frames it the deeper problem with the idea of species has to do with the ineluctable character of real essences. Awareness of the distinction between the real and nominal essence of substances includes implicitly accepting that even observable resemblances are irreducibly provisional, and thus a realistic assessment of natural kinds must tend toward an astonishing proliferation of species: "each abstract idea with a name to it, makes a distinct species" (3.6.38.463). Locke's radically particularist ontology rejects the traditional notion of species because it undermines the conception of real essences upon which natural species depends: "it is absurd to ask, whether a thing really existing, wanted anything else to it" (3.6.5.441). Moreover, as the distinction between kinds of dogs such as a "Shock and a Hound" clearly suggests, nominal species are subject to internal modification through natural processes (3.6.38.463).[9] Simply put, species are artificial categories with no inherent relation to the real essences of natural things.

For Locke, the problem of essentializing species is particularly acute for moral philosophy with respect to efforts to define the human species. Throughout the *Essay* Locke examines a number of definitions of the human being in order to answer the question: Who or what is a human being to whom one has moral obligations? A number of these fairly standard definitions emphasize reason as the defining characteristic of what it means to be human. For instance, Locke raises the possibility that it is the power of abstraction that most clearly distinguishes humans from the "Brutes" (2.11.11.160). He also considers the definition of human as a "corporeal rational creature" or a "Forensic term" applied to a rational being capable of obeying certain moral laws (3.11.16.516; 2.27.26.346). These various definitions share the traditional philosophical premise held since Aristotle that the species human is defined by rationality.

At no point does Locke simply accept the rationalist basis of human identity *tout court.* Instead, he ruthlessly exposes the deeply inegalitarian potentialities of the traditional celebration of reason by claiming that given the considerable range of intellectual capacities within humanity, it can be the basis for a massive system of exclusion. He suggests that typically it is body shape rather than rational faculties that determines membership in the human species. For most people, a severely mentally defective human is "a dull irrational Man," but a bird that can hold a conversation is still just "a very intelligent rational Parrot" (2.27.8.333). Locke's point is not to denigrate the view of the common person vis-à-vis the educated elites. Rather, he suggests that there is some sense in the common view emphasizing body shape. The thrust of Locke's account of species is to show that the philosophical attempt to define natural kinds is much more difficult than typically supposed. The many debates about the moral status of "naturals," "changelings," and fetal monstrosities are, in Locke's view, reflections of the fact that each of these beings with its own "particular constitution" belies the rigid categories of species produced by the focus on real essences (3.6.16-7.448-9). For Locke, from what we do know about the nature of substance, which is of course very little, we can at least say conclusively that rationality does not exhaust the definition of the human.

Locke's treatment of species indicates that our ignorance of substance is a problem for our understanding of morality. The human being is both substance and mixed mode at the same time, and thus with regards to the internal constitution of the human we can only have limited insights. To define the human simply on the basis of *logos* as it has been traditionally understood is to open moral philosophy to a dogmatic and unscientific inegalitarianism. However, to identify the human simply on the basis of external sensible resemblances is also potentially to allow for a massive system of exclusion based on the most superficial features (4.7.16.607). Locke's epistemology thus seems to discredit the whole logic of natural species precisely because there is no way to establish demonstrative certainty about the essence of the human: "None of the definitions of the word Man, which we yet have, nor descriptions of that sort of Animal, are so perfect and exact, as to satisfy a considerate inquisitive person much less to obtain a general consent" (3.6.27.455). Locke does not mean by this that moral philosophy is doomed to intellectual paralysis. His point rather is that terms like species have often been used hitherto to "darken truth and unsettle People's Rights" (3.10.13.497). Presumably the "considerate inquisitive person," who understands the distorting effects of the essentialist tendencies in traditional moral and natural philosophy, will make the first real progress toward enlightenment by recognizing that natural and moral philosophy operate on very different premises.

THE PROBLEM OF DEMONSTRATIVE MORALITY

Locke raises the epistemological stakes in the relation of moral and natural philosophy dramatically when he advances perhaps the most striking claim in the entire *Essay;* namely, that "Morality is capable of demonstration, as well as Mathematicks" (3.11.16.516; 4.3.18.549). Moral ideas are like mathematical propositions in the sense that both involve mixed modes that in theory avoid the problem of intermediation between the idea and the reality of things, precisely because they are arbitrary constructions of the human mind. Locke also observes that the mathematical model is incompatible with the study of existent natural bodies. The elusive real essences of substances leads Locke to suggest that, in contrast to math or morality, "natural Philosophy is not capable of being made a Science" (4.12.10.645). Locke insists that given the limitations of our knowledge of substance, "'tis rational to conclude, that our proper imployment lies in those Enquiries, and in that sort of knowledge, which is most suited to our natural capacities, and carries in it our greatest interest," namely, eternal salvation (4.12.11.646). This is frequently interpreted to mean that Locke contrasted the inevitable mediocrity of natural science with the certainty of moral philosophy.[10] However, it may be wise to avoid jumping to this conclusion too briskly.

When we carefully reconsider Locke's arguments about the possibilities and limitations of natural science and moral theory respectively, we are left with the strong suspicion that Locke believed that moral knowledge is less certain and accessible, and the knowledge of natural bodies more rigorous and expansive, than it first appears in his account. To start, while Locke identifies the acquisition of moral knowledge as the Proper "Business of Mankind," he hastens to add that he "would not therefore be thought to disesteem, or dissuade the study of nature" (4.12.12.647). Morality is the proper business of human beings partly because it is a study suited to an egalitarian distribution of rational capacities, whereas unlocking the "Secrets of Nature" is suited to the "private talent of particular Men" such as the "Master Builders" of modern science like Newton, Boyle and Sydenham (4.12.11.646, "Epistle," 9-10). The elitist tendencies in the study and practice of natural science do not, of course, diminish the very positive contributions made to human flourishing by scientific advances, and Locke recounts with great aplomb the civilizational significance of the discovery of iron and quinine (4.12.11-2.646-7). Knowledge of substance rests on arbitrary nominal essences; however, Locke suggests that the nominal accounts of species are capable of better and worse formulation. This is revealed with delightful irony in his comparison between the power of human senses and that of a microscope when he suggests that if only human beings had microscopes for eyes, we would approach the knowledge of substance presumably enjoyed by angels and other intelligent immaterial species (2.23.11-2.301-03).[11] The irony is, of course, that, as Locke was well aware, human beings are capable of supersensory perception through the invention of technology like the microscope, and in the hands of a trained professional this device can potentially tell us a great deal about the previously unobservable properties of natural substances.[12] Locke herein even subtly implies the possibility of revolutionizing atomic theory with the discovery of ever more minute subatomic particles.

Locke's presumed pessimism about natural philosophy could thus be seen less as a diminution of natural science vis-à-vis moral philosophy, and more likely as a caution that the real promise of natural science not be obscured by the excessive theorizing, and resultant diminution of the importance of experimental science, typical of the scholastic, Cartesian, and Spinozist view of natural philosophy. In both traditional metaphysics and the new *scientia,* Locke identifies totalizing tendencies in their philosophic premises that result either in natural science being absorbed into moral philosophy or moral philosophy being swallowed up by a materialistic naturalism. Philosophy is, for Locke, inextricably connected to allowing and encouraging reason to examine nature, morality, and for that matter theology, on the basis of distinct forms of evidence proper to each endeavor.

In order to maintain separate spheres of knowledge pertaining to moral and natural philosophy, Locke advances the claim that the real problem in understanding the proper relation of natural and moral philosophy lies with the presumptuous metaphysician, rather than the humble experimentalist. Locke stresses that there is no perfect science of bodies, and thus with regard to natural science: "Experience here must teach me, what Reason cannot" (4.3.26-29.556-60). By emphasizing the hypothetical character of natural science, Locke neither precludes better and worse judgments of bodies based on empirical observation, nor does he set an absolute predetermined limit to what conclusions experience can validate. However, Locke does intend to chasten certain proponents of the modern scientific perspective who believe that the conclusions of natural science can be the basis for a new metaphysical dispensation. In calling on metaphysicians to be wary of overconfidence about a complete science of bodies, Locke in no way intends to discourage experimental science. Even corpuscularianism he offers as a working hypothesis rather than a complete science (4.3.16.547). In Locke's view natural philosophy may be an endless task, but it is certainly capable of real and tangible advances in understanding.

In addition to observing Locke's perception of the potential advances of natural science, it is also useful to recognize the important qualifications he places on the certainty of moral philosophy. In a crucial passage in chapter twelve of Book IV of the *Essay,* Locke restates his famous claim about the demonstrativeness of moral knowledge with the major caveat, "if a right Method were taken, a great part of Morality might be made out with that clearness . . . of the Truth of Propositions in Mathematics" (4.12.8.643-44). Not only does certainty in morality depend on use of the "right method" for acquiring knowledge, Locke insists that even then only a "great part," but not all of the rules of morality, can be known with mathematical certainty. The truth of the matter is that in practice morality is not capable of the degree of certainty possible with mathematical propositions.

The epistemological problem produced by language and the difficulties Locke identifies with establishing clear definitions of terms ensures that mathematical truths, unlike morality, need not be "lost in the great wood of words" (4.3.30.561; 2.28.4.351).[13] Were moral philosophy able to replace what Locke rather humorously calls the "prevailing custom" of using words with more clearly defined symbolic representation, it would in theory be possible to approach the certainty of math in a regular way. He even encourages moral philosophers to imitate natural scientists who employ drawings and sketches, rather than words, to record their observations (3.11.25.522-3). Notably Locke does not in the *Essay* lay out in detail the proper method of moral demonstration. However, he does succeed in drawing the reader's attention to the striking parallel between the limitations of moral and natural science. While Locke hardly displays philosophical naiveté, he insists that his epistemology can assist natural and moral philosophy by "clearing Ground a little, and removing some of the Rubbish, that lies in the way to Knowledge;" that is by helping to clarify their terms, sorting through theoretical confusions and making their practitioners aware of the limitations in the way in which the human mind represents the world in thought and language ("Epistle," p. 10). The chief lesson Lockeans can draw from this is to beware overconfidence in the power of words to grasp truth not only about substance but also in moral matters. Only in this sense can the philosophical study of ethics and morality provide them with greater rigor, even as Locke trims back metaphysical claims that have traditionally accompanied these studies.

A NEW CONCEPTION OF PHILOSOPHY

Locke's treatment of moral and natural philosophy presents them in a problematic relation on a number of levels. The *Essay* raises the possibility of demonstrating moral truth, while simultaneously showing the extreme difficulty in doing so through language. It advances a conception of natural science that is both limited by our inevitable ignorance of the real essence of substance, and yet is capable of better and worse hypotheses drawn from the experimental method. For Locke then, the human being is both a rational being capable of knowledge and at the same time a kind of substance whose real essence is ineluctable. How then can we determine who is human or what characteristics are definitive properties of the human species in any but an arbitrary, and hence irreducibly provisional, way?

It is at this point that we need to take up the theological questions raised by Waldron and Zuckert that we set aside earlier. In Book I of the *Essay* Locke suggested "the only true ground" of morality is a God "who sees men in the dark, [and] has in His hands Rewards and Punishments" (1.3.6.19). Thus, the normative conception of the human ultimately seems to be connected to proof of the existence of the divine "Law Maker" (2.28.5.351). Later in Book IV Locke's two central arguments proving the existence of God derive from our intuitive knowledge that nothing comes from nothing and that "incogitative matter" cannot produce a thinking being (4.10.3.620; 4.10.10.623). Apart from the several ambiguities and even theoretical inadequacies contained in it,[14] what is probably most striking about Locke's account of the proof of the existence of God is that it does not, as we might expect, set up a rational demonstration of the particulars of the moral law vouchsafed by God, but rather it introduces a lengthy treatment of the grounds of belief epistemologically falling far short of knowledge. It turns out that humankind's "proper business" of securing salvation has more to do with judgment, probability, and faith—mental activities that do not claim certainty—than it does with demonstrative morality (4.15-14.652-7).

For Locke, reason is central to faith insofar as it involves the mind's assent to the propositions of revelation. Revelation provides insights about the immortality of soul that are unavailable to natural reason, but are never contradictory to it. For our purposes, however, the salient point is Locke's insistence that faith, even when grounded on a high degree of assurance, is still conceptually distinct from knowledge. As he argues: "With what assurance soever of believing, I assent to any article of faith, so that I steadfastly venture my all upon it, it is still but believing. Bring it to certainty, and it ceases to be faith."[15] Locke admits that given that "our knowledge is so narrow," it is inevitable that the "greatest part of Men, if not all" must have diverse opinions about crucial areas of religion and morality without certain proof of truth (4.16.4.659). In this respect theology and natural science are more similar to each other, at least with respect to their epistemic basis and their reliance on probability, than either are to demonstrative moral knowledge.

Zuckert is thus correct to observe that in lieu of compelling evidence for the existence of God and the immortality of soul, Locke's epistemic claims for moral philosophy run the risk of collapsing for all practical purposes into the same epistemological zone as natural philosophy; that is the realm of belief rather than knowledge.[16] Indeed, Locke's concern about this problem may explain his insistence, upon scant proof, that morality is capable of demonstration. However, this difficulty is resolved at least partially if we accept that Locke conceived of demonstrative morality in two senses.[17] First, there is what we could call the weaker version of moral demonstration, according to which Locke emphasizes the certainty that is produced by the perception of agreement between ideas. If the connection between the discrete ideas forming a moral concept such as homicide or adultery can be demonstrated through logical steps by means of "intervening Proofs," then in a sense we are capable of acquiring "real knowledge" of morality, in addition to mathematics (4.2.4.532, 4.4.7.565). However, as is true of any mixed mode this is largely a formal demonstration with no necessary connection to a real pattern or archetype in nature, and thus the normative character of the moral idea remains still somewhat murky. A second sense of moral demonstration would involve a stronger version of the argument. This would require establishing a potentially quite lengthy chain of rational deductions directly connecting the ideas contained in a moral concept to an ultimate source of obligation. This version of moral demonstration would theoretically deal with the problem of normativity by positing a God who rewards and punishes in the afterlife, but this brings us back to the difficulties we have already encountered that are raised by Locke's clear conception of the limits on human understanding. Perhaps the only conclusion we can safely reach on the basis of Locke's argument is the far from satisfying possibility that Locke did not see rational demonstration as the only form of moral reasoning available to human understanding. Otherwise, his claim that we have "light enough" (1.1.5.45) to know our moral duties would be unintelligible.

While it is possible to make the case, as Zuckert persuasively does, that Locke complemented his theological argument for morality with a secular argument rooted in his theory of self-conscious personal identity,[18] the present study seeks to highlight a different aspect of Locke's conceptual flexibility and suggest that Locke's primary aim in the *Essay* was not to develop his moral theory, but rather to outline a new conception of philosophy based on his epistemology. The combination of Locke's epistemology and particularist ontology has the effect of undermining all manner of metaphysical philosophy. "Ground clearing" militates against the ambitious metaphysical system-building in both its classical scholastic and modern materialist forms. Locke limns the features of a new and more modest conception of philosophy. His effort to narrow the definition of what constitutes knowledge and his related attempt to broaden our sense of the subjective intellectual processes involved in belief formation does indeed lay parallel secular and theological tracks running through Locke's moral philosophy. However, Locke's goal in the *Essay* is not to advance a fully fledged moral philosophy, but rather the more limited aim to establish new parameters for philosophic activity framed within a cognitive range bounded by substance, on the one hand, and the supernatural, on the other; between the mysterious properties of the blade of grass and the incomprehensible operation of the "eternal infinite Mind" of

Almighty God (4.10.19.630). In the complex double-movement of Locke's philosophy both substance and supernature are simultaneously rationalized and problematized.

Arguably prior to Locke, philosophy was by definition a comprehensive and synthetic intellectual activity encompassing mathematics, physics, and theology as well as logic and ethics. In the *Essay* we see perhaps the genesis of modern analytical philosophy in Locke's reduction of the scope of philosophy to its proper "underlaborer" role in clarifying terms and concepts in service of philosophy of mind, language and political philosophy. By making philosophy conceptually more compact, Locke seeks to do two things: (1) focus on determining sound methods and procedures for the operation of mind, and (2) prevent both natural science and theology from simply being absorbed into the rigid categories of metaphysical philosophy. Locke's understanding of the corrective, or as Rogers describes it "therapeutic," function of philosophy involves then freeing experimental natural science from unscientific dogma and moral strictures, while also liberating moral philosophy from the materialistic and mechanistic tendencies of scientific naturalism.[19] In the process Locke carves out an epistemological zone of relative freedom for philosophy ideally to support intellectual activity embracing both the possibility of certainty and the likelihood of probability.

The impact of Locke's new direction for modern philosophy on the political realm is obviously considerable. In his account of the relation between natural and moral philosophy, Locke indicates the great difficulty in trying to isolate the moral idea of the human from the experience of being human substance. But where does this leave us with respect to the political issues so central to Waldron and Zuckert?

Clearly on one level, Locke sets natural and moral phenomena as separate spheres of study with their own forms of evidence and claims to truth. The uneasy balance between natural science and moral philosophy is what I would call Locke's "wager" regarding the divided mind of liberal society. From the skeptical tone of the *Essay* and evidence we can glean from other works in his corpus, it is perhaps fair to say that Locke anticipated that liberal society would be animated by a moderate and pragmatic spirit with regards to both natural and moral philosophy. The habitual conservatism of a science produced by the humble experimentalist and the rational morality maintained by followers of Locke's "reasonable" Christianity do appear to be ideal partners. Moreover, the stunning success of liberalism since the end of the Cold War appears to testify to the political viability and indeed durability of this balance. However, is there not, at least on a theoretical level, a deep structural instability in this arrangement as well? A natural science constitutionally incapable of endorsing, or even inclined toward constantly refuting, the concept of natural species could, as Waldron fears, seep into thinking about the human species and thus undermine liberal society's confidence in the natural basis of moral and political equality.

Conversely, a new moral philosophy made aware by Lockean epistemology of the arbitrary character of its own modal reasoning may have few intellectual scruples about simply dismissing or discrediting empirical evidence produced by natural science that challenges regnant moral principles. Needless to say, probability does not exactly inspire automatic authority. As recent developments in some liberal societies have shown, perhaps it was predictable that the divided mind of liberalism would reflexively take a "Red State/Blue State" political expression with factions moral and religious, on the one hand, and scientific and secular, on the other, doing battle over the very meaning of what constitutes evidence, knowledge, and proof.

The real source of the problem appears to lie in the fact that by limiting the scope of knowledge and vastly expanding the range of belief, Locke's epistemology accentuates the political relevance of moral ideas, or for that matter any idea, which can in principle be clearly defined and tested through rational demonstration. Like any commodity, knowledge is made more precious by its scarcity. On one level, Locke seems to obviate the possibility for clashing claims to knowledge emerging from moral and natural philosophy by reconceptualizing them both primarily in terms of probability. However, their balance in what he calls the epistemological "twilight" realm of probability may not be as stable as Locke presents it. To start, while he insists that reason is operative in both moral and natural philosophy, the conclusions reached on matters of faith, and by extension morality, do not, unlike natural philosophy, expect confirmation by observable empirical evidence. This is not to suggest that there is no conceivable empirical basis to evaluate the conclusions of moral reason, but rather that, again unlike with natural science, Locke did not explain how we can produce or expect rational agreement about what this empirical basis would be. As Locke was well aware, in some sense moral disagreement and debate is always about the theoretical criteria by which to judge practical experience.

A second aspect of this problem emerges in Locke's account of a kind of knowledge about "the particular existence of finite beings without us," which he termed sensitive knowledge and insisted relies much more heavily on experience than intuitive or demonstrative knowledge (4.2.14.537-8). For this reason it is the weakest kind of knowledge, but Locke insists that sensitive knowledge has a claim to certainty, and is thus epistemologically speaking of a higher order than probability. Locke perhaps signals his concern about this in his educational treatise *Some Thoughts Concerning Education* when he suggests that students be introduced to religion and metaphysics prior to the study of physical science. His reasoning is that this is "good preparation for the study of bodies," not because it encourages any specific moral behavior, but rather inasmuch as developing in the student a sense of the possibility for the existence of spiritual beings may check the potential excesses of the materialist bent natural to the human mind, which is of course supplied with much of the material for knowledge by the senses.[20] The point to which Locke appeared peculiarly insouciant is that while the stakes are certainly higher for morality and religion than for natural science, the epistemic grounds of the former modes of reasoning also appears to be weaker. It may be more important to understand our moral duties and our potential chances for salvation than it is to learn about the physiology of the koala bear; but the fact is that on Lockean terms we can probably *know* more about the koala than we ever can about salvation and possibly even of our moral duties.

The tension or conflict between natural and moral philosophy thus arises from a condition in which the claim to know something about less important matters can logically lead to substantive claims about more important matters like the biological basis for human species membership—something with obvious moral and political implications. Natural philosophy as Locke conceives it has the epistemic means to slip the leash of probability and make plausible claims to knowledge with greater facility than is the case with either morality or faith. For its part, Locke's idea of faith reduces reason to a largely instrumental role in the validation of the miraculous, and thus may not be as well equipped as Locke thinks it is to harness the supernatural in support of a rational system of ethics. As Locke was fully convinced, either in response to scientific intrusions into matters of morality and faith or as a separate development, religion is quite capable of transgressing its rational constraints by falling back on mysticism or scriptural literalism.[21] Both science and religion have the means in Locke's system not only to defend themselves, but also to attack each other.

Supposing that Locke recognized this possibility, is his new conception of philosophy fit to play a role in preserving the uneasy balance between natural and moral philosophy? In philosophy's traditional role as the master science, it was conceptually well equipped to be the mediator among conflicting claims to knowledge by resolving, harmonizing, or simply bracketing off disputes involving theology, ethics, and natural science. Admittedly it did so often at great cost to theoretical coherence, but at least prior to Locke there was a general presumption that this is precisely what philosophy is supposed to do. The great danger to which Locke may not have been sufficiently attuned is that his more modest, and in some respects diminished, idea of philosophy no longer has the epistemological means to make authoritative claims to knowledge about what is human or contributes to human flourishing. In this sense, the tension between the scientific and nonscientific modes of Locke's conception of reason is potentially destructive because it is unmediated by philosophy. Enlightenment optimism about the natural convergence of moral and scientific knowledge perhaps presupposed a more ambitious idea of philosophy than Locke was willing or able to articulate and defend. Contemporary philosophy is sometimes criticized for not being able to speak to the political and scientific communities in their own language and in terms of commonly recognized theoretical first principles. Often this is a fair criticism that was leveled against Locke himself.[22] More problematic, however, may be Locke's apparent inability to conceive of philosophy as a means through which the moral and the scientific aspirations of modernity can meaningfully speak to each other.

ENDNOTES

1. Michael Waldron, *God, Locke and Equality: Christian Foundations in Locke's Political Thought* (Cambridge: Cambridge University Press, 2002), chapter 3.

2. John Locke, *Two Treatises of* Government. Peter Laslett, ed. (Cambridge: Cambridge University Press, 1988), Second Treatise, section 4. See also Second Treatise, section 128 and First Treatise, sections 30–31.

3. Michael Zuckert, "Locke—Religion—Equality," *The Review of Politics 67,* no. 3 (Summer 2005): 419–31.

4. John Locke, *An Essay Concerning Human Understanding,* Peter Nidditch, ed. (Oxford: Oxford University Press, 1975), book 1, chapter 3, section 22, page 81; bk 2. ch 1.sec 2. p 104, bk 1. ch 2. sec 15. p55). Hereafter in notes and text bk, ch, sec, and pg.

5. For Locke's attack on Aristotle's concept of real essences, see Michael Ayers's excellent discussion (*Locke: Epistemology and Ontology,* [New York: Routledge, 1993], Vol. II, 70–74).

6. Ruth W. Grant, *John Locke's Liberalism*, (Chicago: University of Chicago Press, 1987), 16.

7. As Rogers observes, this is the basis for Locke's rejection of Cartesian dualism as well as scholastic metaphysics (G.A.J. Rogers, "The Intellectual Setting and Aim of the *Essay,*" in *The Cambridge Companion to Locke's 'Essay Concerning Human Understanding'* [Cambridge: Cambridge University Press, 2007]), 7–32, esp. 14.

8. Ayers, *Locke,* 68.

9. In this respect Locke's idea about the fluidity and inherent instability in the concept of species clearly foreshadows Darwin's account, although Locke does not of course identify anything like the notion of natural selection.

10. See for example John Colman, *John Locke's Moral Philosophy* (Edinburgh: Edinburgh University Press, 1983), 4–5 and Catherine Wilson, "The Moral

Epistemology of Locke's Essay" in *The Cambridge Companion to Locke's "Essay Concerning Human Understanding"* (Cambridge: Cambridge University Press, 2007), 381–405, esp. 382.

11. John Yolton suggests that Locke entertained the possibility that there is an internal structure of physical bodies from which we could, in due course, build a deductive science of nature (*The Two Intellectual Worlds of John Locke* [Ithaca: Cornell University Press, 2004], 27).

12. Locke was, of course, intimately familiar with the work of the great pioneers of microscopy, Anton von Leeuwenhoek and Robert Hooke, who were acquiring much attention in scientific circles in Locke's time. For Locke's contact with Hooke and Leeuwenhoek, see Roger Woolhouse, *Locke: A Biography,* (Cambridge: Cambridge University Press, 2007), 158, 164, 239.

13. For an interesting discussion of the problematic effect of Locke's linguistic theory on his account of civil society more generally, see Hannah Dawson, "Locke on Language in (Civil) Society," *History of Political Thought,* 26, no. 3 (Autumn 2005): 397–425.

14. See Nicholas Jolley, "Locke on Faith and Reason," in *The Cambridge Companion to Locke's "Essay Concerning Human Understanding.*" Lex Newman, ed. (Cambridge: Cambridge University Press, 2007), 436–55, esp. 439.

15. John Locke, "Reply to the Bishop of Worcester's Answer to his Letter," *Works, Vol. IV* (London: Tegg, 1823), 146–47.

16. Zuckert, "Locke—Religion—Equality," 429–30.

17. For thoughtful discussions about the ambiguity in Locke's conception of demonstrative morality, see Greg Forster, *John Locke's Politics of Moral Consensus* (Cambridge: Cambridge University Press, 2005), 97–100 and Nicholas Wolterstorff, *John Locke and the Ethics of Belief* (Cambridge: Cambridge University Press, 1996), 142–48.

18. See especially, Michael Zuckert, *Natural Rights and the New Republicanism* (Princeton: Princeton University Press, 1994), chapter 9.

19. Rogers, "Intellectual Setting," 30.

20. See John Locke, *Some Thoughts Concerning Education*, Ruth Grant and Nathan Tarcov, eds (Indianapolis: Hackett Publishing, 1996), section 190–92. Compare this with Locke's *Essay* 2.1.2-3.104-05.

21. In the *Essay* Locke identified the problem of religion divorced from rationalism as "Enthusiasm" (see bk 4. ch 19).

22. See for example, Locke's exchanges with Jonas Proast and Edward Stillingfleet, the Bishop of Worcester, over the issues of toleration and rational theology respectively (John Locke, "Third Letter Concerning Toleration," *Works, Vol. VI* [London: Tegg, 1823], 59–574 and "Reply to the Second Letter of the Bishop of Worcester," *Works, Vol. IV* [London: Tegg, 1823], 193–498).

Chapter 14

Rousseau's Botanical-Political Problem: On the Nature of Nature and Political Philosophy

Leonard R. Sorenson

WHOLE OR ALL?

Our theme is the meaning of and relation between modern reductionist science and political philosophy. The question is whether political philosophy can meet the challenge of reductionist science and retain or regain its place as the queen of the social sciences. The answer turns on the adequacy of reductionist science to account for all that is real, including life, or whether recourse to forms or wholes is necessary to account for the beings.

Rousseau's moral critique of modern reductionist science or reason is by now commonplace. It is linked to vice in its origin, ends, and effects. In particular, it is associated with pride, and it ultimately undermines human good or happiness. Yet, Rousseau himself was a top-notch natural scientist. He produced a very impressive body of work on chemistry, and especially botany.[1] As will be shown, exploration of the latter works reveals his philosophical critique of modern reductionist science as unable to access important truths about beings.

In his botanical works, Rousseau critiques reductionist reasoning on a very comprehensive level. He moves to his subject matter, plants, by means of division and collection, according to the natural kinds of beings. As an "observer," not a "moralist," he gives us to understand that we live as a part of a complex or heterogeneous whole. All is in truth a whole that entails naturally irreducible differences in kind: the nonliving, "combinations of compounds;" plants, partaking of "spontaneous," nonlocal "motion;" animals, "spontaneous, progressive, feeling and seeking" beings; and the human animal, the "sentient, active, seeking and thinking being." Thus, we are and live in neither an infinitely differentiated all nor a homogeneous whole, neither a many nor a one. Moreover, we live in a heterogeneous whole that entails irreducible parts which, in turn, are wholes with irreducible parts, and so on, down to "simples" which are also ultimately wholes with parts, complex "combinations of compounds."[2] Two points concern us: first, the place of whole—part or form—matter in Rousseau's natural science; and second, the evidently *natural* status of man as the seeking, thinking being.

NATURE AND THE "GENERATION OF PLANTS"

As we move to Rousseau's study of plants it will prove useful to remind ourselves, first, of his more famous claims about nature, and, second, of some additional introductory statements found in his botanical works. In addition to his claims that nature—on the whole—is good, and does not lie, and that man is naturally good—we find the following general statement in his *Discourse on Inequality among Men:* "Man gets the idea of how nature proceeds from the study of the generation of plants." Then in his botanical works we find the following general statements. First, "nature does nothing is vain." Second, "nature has provided a summary of her work" in plants, especially their flowers. Lastly, plants exhibit an "irrevocable" law, the "law" of the "nature" of all embodied beings.[3] As we will see, the noteworthy points of these statements are, first, that one can learn something important about the nature of nature as such, including the nature of man, from the study of plants; and the second point, in the form of a question, is what is the law of nature and its relation to modern reductionist science?

What is a Plant?

As with his account of the natural whole, Rousseau begins his "Introduction" to botany with a serious qualification of reductionist science. Without naming names (Bacon and Descartes), Rousseau critiques "certainty" as the very meaning of possible knowledge, as well as its goal, the ultimately instrumental purposes of certainty as knowledge, namely, to "make the entire human species immortal." To "regard" plants in this way is to study their "virtues," that is, their relation to "effects on the human body" for the sake of physical cures. This requires, in turn, the reductionist study of their "matter," their "simples"—not as the complex wholes or as "compounds" in "combinations"—but to find derivative "remedies" to cure sickness and even death itself. Botany is thus reduced to instrumental chemistry and physics which abstract from the phenomena itself: both the plant itself and its matter or simples as ultimately compounds or complex wholes with effects greater than the sum of the parts.[4]

However, Rousseau, as natural scientist, as an "observer" not a "moralist," as a botanist, asked the paradigmatic Socratic question: what is a plant? As such, he sought "knowledge of plants themselves," even "knowledge of a single plant." As their beauty is but an "invitation to study," the latter, not the former, is the "best half" of botany. He seeks to "exhaust" a single plant and even all plants, that is, to come to know all that is knowable about a plant or plants. Since plants are wholly natural, he seeks to discover the very nature of each and all plants. Since each plant *is* at least as much *this* plant as it is this *kind* of plant, there is much to learn.[5] Fortunately, it is adequate to our purposes, to move from general to more specific, and from negative to more positive, accounts in order to discover Rousseau's understanding of the nature of nature.

Plant: "Heterogeneous Substance"

To begin with, a plant is neither an "absolute substance" nor merely epiphenomena of its aggregate parts, its "matter" or "simples." Nothing is either formless matter or matterless form: Rousseau's word is "heterogeneous substance," or whole.[6] To be a bit more specific, the purposes or ends of a plant are self-reproduction and death. Its movement—birth, growth or development—is oriented to and by these ends. Plants—and all life forms—exhibit the life cycle: birth, growth, self-reproduction, maturity, decay, and death. This is an "irrevocable law," the "law of (the) nature," of all embodied beings.[7] Thus we begin to see the law of the nature of irreducible complex beings. More specifically, the plant forms or organizes itself into its shape, its parts, and the structure of its parts, to perform operations for its ends. It "brings together" its parts which are also subwholes with parts. They mutually influence each other just as the whole forms and informs them. A plant is a whole and parts, a whole greater than, with effects on, its parts. It transforms nonliving beings into its organism.[8] More generally, a plant entails spontaneous internal active causation combined with a passive, receptive, or responsive capacity in relation to external causes.

Moreover, that which is referred to as "life," the "principle of life," causes multiple, diverse, structured change or movements toward ends. The plant life form is thus a heterogeneous whole, a complex unity of form and matter, an irreducible emergent complexity, a self-emergent whole. In all, a plant exhibits the marvelous mystery of "metamorphosis."[9] For our present purposes, an important truth to extract from the "irrevocable law" of the nature of plants is precisely this: structured movement or change as activity; and activity as the process of actualizing natural potential.

Plants and the Human Mind or Soul

Now, since "we get the idea of how nature proceeds from the study of the generation of plants," it follows that the natural way to study is also exhibited by the proper way to study plants. Rousseau thus proceeds in his botanical works to exhibit his own education and to teach others the proper, natural, way of learning.[10] As will be shown, these learning activities constitute actualizing the natural potentials of the soul and mind of the human being.

Since the human animal is the spontaneous, progressive, seeking, sentimental and thinking being, it follows that man must be in need of learning, of education. Man must learn what "suits" him.[11] For instance, though plants are generally a delight to eye and good to eat (figs, see the Garden of Eden), some are beautiful but not good (hemlock, see Socrates), some are good but not beautiful; and some are good but taste bad. Man must learn the distinction between the naturally good and the naturally beautiful for man. Though all that is natural is on the whole good, only some of the natural is good for man. And not all that is beautiful is good for man. In short, man must learn to know the distinction between the good and the beautiful in the search for what is truly good for man.[12]

More interesting is a more elevated need and hence stance toward plants in relation to actualizing natural potential. Rousseau describes the natural movement or change from little or no awareness of a plant to progress in knowledge of the "true character" of a plant, the "natural notion" of the plant.[13] As a given plant "presents itself naturally," to display or "show [its] visage," its natural to-be-noticeables, its surface, shape and therefore colors; so man can only come to notice plants or a plant in the way man is most naturally able to notice, namely, by a sort of intellectual glance, a natural power of the mind's-eye.[14] To fully appreciate this initial awareness—intellectual perception—one must be moved by "disinterested love," love of nature" and hence "truth." One must "learn to love nature," to develop a "healthy," as opposed to a "furtive," mind, and learn "sound reasoning" based on "careful observations."[15] These, in turn, constitute the activity of actualizing natural potential. For instance, as mind and sight are linked, so both must be mutually "cultivated" by "practice," by "exercise" in order to learn *how* to see and to know "little by little" by trial and error, a plant or plants.[16]

More specifically, in the beginning is the look, the surface-shape-form and hence color of the plant. It is a sort of unity or whole that is this-not-that-being, a thing—and this kind of—being. Its dimensionality, unity or solidarity, is intellectually perceived by man as a sort of primal solid geometry. Man notices its surface-shape-form-color as its unity or wholeness. On the basis of this initial movement to and from awareness of a plant, to the beginning of knowledge of a plant, man proceeds by "analysis."[17] This is the movement to the parts that are as such of the whole, to the mutual relation of the parts and back to the whole that constitutes the parts. The goal of this dialectical reasoning is to "exhaust" the plant, to know all that is knowable about the plant, including the purpose of each part in relation to each other and the whole. The limit, the least incomplete knowledge, in Rousseau's view, is clear. On the one hand, is the aforementioned life cycle, the law of the nature of embodied beings. On the other hand, is the mystery of what is called the "principle of life." For life is both one, "unity," but also is with diverse parts or functions, such as nutrition and growth, etc.[18]

Moreover, analysis or "sound reasoning," dialectical reasoning, based on careful observation begins and ends—exists between—two mysteries: one, on the side of man—intellectual perception—and two, on the side of the plant—the marvelous mystery of metamorphosis.[19] The proper exploration of the natural potentials of the plant constitutes simultaneously the actualizing of the natural potentials of man.

NATURAL POTENTIALS OF MIND OR SOUL

A major strand of Rousseau's monumental study of man, his *Discourse on Inequality among Men,* prefigures his account of man in relation to nature, including plants, in his botanical works. As we will see, it also constitutes the themes or problems upon which political philosophy is oriented.[20]

The unity of the two works is the relation between human nature and "soul or mind." In the beginning of his *Discourse on Inequality among Men,* Rousseau introduces a truth about human nature that takes a trajectory throughout the work. That truth is that there is "natural or physical inequality in mind or soul" among men. These natural inequalities of "qualities" are called "strengths," especially, "powers." As in his later botanical works, Rousseau casts his account in the language of natural potentials.[21]

More specifically, Rousseau develops and details this truth of natural inequality over the body of work. Natural inequality in mind or soul is coequal with, but "nascent" in, man. From the very beginning, we always "had" it, naturally "received" it, and it "was" always a "capability." Its "reality" now is "far" from what it "was." What is more, these natural potentials or faculties have a "limit" or "perfection." The "perfection of the individual" is the "perfection of reason" and the perfection of reason is dialectic. The human mind or soul is not simply malleable. The range of this natural inequality, when "cultivated," is on the order of the image of the natural "giant" to the "dwarf." Moreover, in the language Rousseau applied later to plants and man, natural human potentials of mind or soul—if properly "cultivated"—can "engender(ed)," "grow active," "unfold," "take shape," "develop," and even "bloom."[22]

Again, as plants are the product of the mutual influence of internal and external causation, so the human soul or mind is subject to both "self-perfection" and "perfectibility." Also, as in his later botanical works, the effects of these causes can be mixed and turn on the role of "amour-propre." Properly subordinated and well ordered the results can be "good" (or the "best") men. Otherwise, men become "bad" (the "worst") men. At best, the perfection of mind or soul is the "honor of the human race."[23]

Furthermore, as certain plants are the best expression of their kind, so the "good" or "best" includes the "giant" among men. In fact, as the "whole world" is "destined for" man, so the human species is for its peaks: the "enlightenment and happiness of the celestial intelligences." Such "intelligences" are the "sublime geniuses" of the human race: Socrates, Plato, as well as Bacon, Descartes, Newton, and a few others. These human beings "deserve all our admiration," and are the "light and honor of the human race." Rousseau "communes" with the products of their minds in reverie. He counts himself among them. He transcends most of them and has few, if any, equals. More generally, these humans are godlike in relation to the natural heterogeneous whole, since the one who deserves the name God would be that "supreme intelligence that knows everything in a single glance." But the peak proper is knowledge of man, and in particular, progress in self-knowledge and of oneself in relation to the whole and among men.[24]

However, since the human life form, the human being, is the spontaneous, sentient, intelligent, seeking being—the embodied soul-mind—it certainly follows that no man is "pure mind." Man too is a complex life form, made by nature to seek what "suits" him.[25] Now, humans are also "persons," that is, able to exhibit a "persona," a "mask." To the disinterested observer, the spectacle of humanity, the civilized mass of humanity, are intellectual "dwarfs" who hide their desires and inferiority behind a "mask." We are such that we are able to double ourselves: be who we think we are and present ourselves as we want to appear to others. We are able to lie, to deceive and be deceived by others and even ourselves. For instance, when one hears another's understanding of the whole, one is typically in a position to learn more about the soul, their soul, than the whole.[26]

Now, behind and beneath the "mask" of man typically lies a secret in the depth of the soul: tyrannical desires, pleasures, envy, jealousy, cruel hatreds.

Philosophers or scientists, even of nature itself, are far from immune from these heinous, turbulent passions. The root of this "poison" in the soul is "amour-propre," pride in relation to fear of death. This is the passion for distinction, to be and be recognized as, or not be but nevertheless desire to be recognized as, the superior, even godlike, human being.[27]

The point is that Rousseau's restoration of the "forgotten education" in proper natural science, modeled on Botany, relates as much to the cultivation of a healthy "soul" as it does to a "healthy mind": attitude and intellect are inextricably—and perhaps inexplicably—mixed in the psyche of men.[28] Hence, the proper natural education in philosophical natural science is the most "useful" for the "wisdom and virtue" of men. Education can be the best "cure" for tyranny. This education can activate wisdom and virtue, as it cures ignorance, error, and vice. Vice and ignorance converge in the psyche—soul-mind—of the human life form.[29] The key at core of the cure is to subordinate "amour-propre" to love, lack or need, the true root of the human being as the seeking being, as the seeker for the suitable for man, the good for man.

Proper education in natural science links love to good and good to truth and beauty. This circumstance can then activate admiration in the soul. Evidently, admiration can outface its contrary—"amour-propre"—and subordinate it to the quest for human good, truth, as well as beauty, that which is suitable for man.[30]

The error of early modern reductionist science, ignorance of form or whole, presupposes that nature including human nature is neutral or hostile to man. As such, nature is unworthy of loving admiration. The truth is that nature is on the whole good, hence worthy of seeking, of loving, of admiration. We live in a heterogeneous, ordered whole in which man is a part. Man is the only part of the whole aware of the whole. Man's part properly constitutes pride of place in the whole.[31]

Yet, the natural potentials of man are a product of nature not man, and man is but a part of the natural whole which, on the whole, is good. Nature deserves the name—God—not man. Hence, pride of place as part of the whole is to be subordinated to love, love of God, love of nature as good. Loving admiration leads to a human stance of conformation to, not the conquering of, nature. Moreover, loving admiration of nature as good, including admiration of oneself in the proper dependent relation to nature, detaches one from dependence upon man, on the opinion of man, and hence dissolves the need or importance of the "mask."

This loving-seeking-admiring-conforming attitude constitutes the healthy "soul" linked to the "healthy mind" in the psyche of man. The proper functioning of the mind and soul together in the activity of true natural science constitutes the actualizing of the natural potentials of man. To be moved to conform means to form oneself towards one's natural form and place in the whole.

As we will now see, Rousseau in his botanical works presents three expressions of this fulfilling activity: contemplative man, primary oriented to beauty; comprehensive man, primarily oriented to knowledge; and Rousseau himself, primarily oriented to knowledge of man, and especially, self-knowledge.

PLANTS: A CURE FOR TYRANNY

To transform tyrannical desire into contemplation of nature "is to chain up love with bonds of flowers."[32] The chain-link bonds two sides: man and plants. On the side of man is whim, fancy, pleasure. On the side of plants are "all the rewards they offer the senses," from mixed to "pure pleasure."[33] The link between the two constitutes a "taste for beautiful knowledge."[34] This experience can put the passions "off" their track and simultaneously activate the "precious serenity of soul" that can arise from "the contemplation of marvels that surround us."[35] Self-sufficiency and pure pleasure converge. As the self is detached from vice, so it is attached to nature through beauty. This trajectory peaks in all but the loss of self. This is the attempt to unite with the harmony of the heterogeneous whole. This experience is a sort of natural pseudo-mystical beatific vision.

Moreover, this trajectory begins with the intellectual perception of the form-unity-whole of the plants at the stage of the life-cycle when they are most beautiful: thus the experience of "beautiful-knowledge." This experience elicits loving admiration because beauty represents the self-fulfilled sufficiency of our natural potential. Man is a complex whole who seeks to be whole, or assimilate into the whole. The loving admiring, seeking conformation to what suits man culminates in all but the dissolution of the self, the very seat of experience, as such.[36]

But "beautiful-knowledge" is not the whole truth. Since beauty is but an invitation to study, and the best half of botany is study, another higher realm opens up for man as seeker.[37] The beautiful knowledge of the harmony of the heterogeneous whole masks the conflict and strife requisite to the equilibrium that preserves the whole. Many truths important to man are not beautiful. This trajectory—like the first—begins in intellectual perception, trust in the senses, especially the eyes, and ends in mystery, the demonstrable limit of knowledge itself: the admirable impenetrables by the mind of man. This is the path or way of the "true botanist" and all philosophical natural scientists.[38]

The crucial difference between the first and the second way is that for the latter "analysis" follows immediately upon the heels of the intellectual perception of beauty. The invitation to study opens up, on the one hand, into the attempt to classify all the parts of entire plant realm: classificatory thinking emerges.[39] On the other hand, there emerge dialectical thinkers oriented on the analysis of the part-whole relation of each plant, right down to the minutest visible part of any given plant. The goal is to wholly "exhaust"—know all that is knowable—about each and all plants. This same mode of study, of seeking, then extends to the heterogeneous whole itself. In both cases one moves from visible to invisible, to mysteries: life hence death, metamorphoses; that which brings and maintains the heterogeneous whole; how form and matter are a unity, etc.[40]

The human wisdom and virtue associated with this quest are considerably different from the first. For instance, in this case, "wisdom" is love of learning, progress in knowledge, and this wisdom, is "crazy," all but daemonic: it dominates, subjugates, enslaves the self under its sway.[41] Love of knowledge of the knowable truth is the tyrant over the self. True lovers of truth are closer to tyrants than to contemplators of beauty. Their tyrannical passions are not exactly "put off the track." Yet, this true lover is neither a conqueror, nor a contemplator, of nature. Rather, this human seeker-lover longs to conform the mind to being, to "exhaust" the beings, to extract all the knowable from being, the heterogeneous whole itself.

This healthy dialectical "mind" is also associated with certain virtues of the "soul," much different from those of contemplative man, and that converge on those of self-knowing man. Comprehensive man develops, nourishes and is nourished by these major "soul" traits: philosophical courage, shamelessness, transcendence of the repugnant and the repulsive, along with laughter.[42]

Moreover, like contemplative man, comprehensive man subordinates "amour-propre" to love, but the latter finds self-admiration in progress in knowledge. And like the former, the latter stands in loving admiration of nature, but in this case, of its mysterious character. Lastly, both are in love with nature, dependant on nature, and are self-sufficient in relation to man, the opinions of mankind: the one oriented to beauty, the other to progress in knowledge of nature.

There is more to the story of Jean Jacques for he, above all, exhibits himself to the reader of his botanical works as engaged in the activity of progress in self-knowledge—the ultimate goal of Botany itself—in the very midst of his own progress in learning about plants. For instance, in teaching the reader the law of the nature of plants, of all embodied life forms—the life cycle—Rousseau exhibits himself as self-aware of his place in the phase of his life-cycle.[43] Over and above beauty, even over and above progress in knowledge of nonhuman nature, stands knowledge of what is knowable about man and especially self-knowledge, knowledge of one's own good and place among men and in the whole.

ROUSSEAU'S "SAD" SYSTEM: POLITICAL PHILOSOPHY

Though nature—on the whole—is good, to conform one's soul-mind to nature requires that one be—to a certain extent—somewhat "sad."[44] Natural inequality in mind or soul constitutes the foundation of political-philosophy proper. In what follows, I draw the link between these two claims.

The highest human activity of "self-perfection" is actualizing one's natural potential for self-knowledge. Most of us, most of the time, are limited by time, circumstance, and—above all—natural potential, to make significant progress in this quest. Political societies, not nature, must be our home. Most of us are informed and formed much more by "perfectibility," than by "self-perfection." What is more, just as the most real in the plant realm is the individual, particular, unique expression of the kind, so it is among human life forms. As complex embodied-soul-mind in relation to complex external causes, the reality of each human being is at least as much this-here-unique-individual-particular human as this kind: human being.[45] However, the most powerful causal agent of perfectibility is political society. Since most of us are intellectual "dwarfs," political society as such must always and everywhere be rooted in categorical conclusion, in prejudice or prejudgment, in opinion oriented to deeds, in the equivalent of Plato's cave.[46] Humanity, like matter, in its own way, is a politically stamped or ramified mass: "sad," but true.

Thus we see the emergence of the first and highest purpose of political philosophy: the quest for self-knowledge. Precisely because it is hard to say how far the deepest opinions of one's age are rooted in one's soul or mind, the would-be political philosopher must work with ruthless dialectic through the layers of the most dearly held opinions that shape our way. One point is clear: these attitudinal-intellectual notions run deep.[47] The life-long quest of progress in self-knowledge is political-philosophy.

The political-philosopher thus discovers along the way the ineradicable, fundamental problem: the disproportion between the good of the individual and the good of the community: the political is also a heterogeneous whole.

On this basis, a second purpose of political philosophy may arise. As depended on, transcending, and in tension with political society, the political philosopher may propose an inherently unstable reconciliation between giants and dwarfs, some link between wisdom and power, the best practical solution for the given times and circumstances.[48]

Thus we see that political philosophy is inherently radical. For instance, who can doubt that Rousseau, in his circumstances, was not fully aware that his newer and truer natural science, especially his botany, as counter-enlightenment-enlightenment—if successful—would prepare public opinion for a revolution precisely by undermining early modern natural science, a fundamental pillar of the emerging politics, rooted in Hobbes and Locke? Again, who could doubt that Rousseau was well aware that his new natural science was also in tension with his own new political society, formulated in his *Social Contract*?[49] The ineradicable truth of the tension between the individual and common good must be recognized as such, even in the attempt of political philosophy to manage it.

Again, Rousseau was certainly aware that as true as it may be that we live as a part of the heterogeneous whole, so it is that a range of human beings—from the properly directed "vivacity" of youth, to the contemplators, on up to comprehensive man, and ultimately to Rousseau himself, as self-knowing man—can come to constitute a formidable fence against the emergence and influence of tyrannical desires in any, even his own, political society. Philosophy is a cure for tyranny.[50] Political philosophy cannot ignore the political bearing of humanity's understanding of its relation to the world or whole.

Moreover, political philosophy proper stands or falls on nature. Neither reductionist nor gracified nature can be the foundation of political philosophy. Political philosophy stands or falls on the idea of heterogeneous whole, and hence of natural human potential.[51]

Rousseau, as a political philosopher, as an embodiment of an answer to the question, what is political philosophy, is certainly open to questions by us as students of political philosophy. But we today, in the grip of the reductionist scientists' claim to hegemony over the very meaning of reason and hence of knowledge, over the answer to the question, what is man, may find Rousseau a worthy entrance into the exploration of the question of the necessity of the fundamental problems that constitute political philosophy as the queen of the social sciences.

*I thank the Earhart Foundation for their generous support of this exploration.

ENDNOTES

1. Citations are from Kelly, C., *The Collected Works of Rousseau.* VIII: 91–255 (University Press of New England, 2000) and XII (2000). We owe a great debt of gratitude to Alexandra Cook for her meticulous collection and translation of

Rousseau's botanical works. See also her, "The Septieme Promenade of the Reveries: A Peculiar Account of Rousseau's Botany," *Studies on Voltaire and the Eighteenth Century,* 2008, n. 3: 11–36. *For* Rousseau's works on Chemistry, see Giuseppe Del Re, "Rousseau and Chemistry," *Hyle-International Journal for Philosophy of Chemistry*, Vol. 7, n. 2, 2001.

2. Kelly, *The Collected Works of Rousseau*, XII: 36, VIII: 93, 96, 99, 108, 109, 115, 124, 128, 130, 133, 137, 139, 140, 156, 174, 191, 215, 251.

3· Gourevitch, V., *Discourses and other Early Political Writings.* (Cambridge University Press, 1997) 168: 22; Kelly, *The Collected Works of Rousseau*, 93, 212. Compare 168, 303 n, 25.

4. Kelly, *The Collected Works of Rousseau*, 93, 212. Neither Cantor nor Kuhn mine the notion of form or whole as cause—and hence natural potential—available in Rousseau's botanical works, Paul Cantor, "The Metaphysics of Botany: Rousseau and the New Criticism of Plants," *Southwest Review*; Summer, 1985, Vol. 70, Issue, 3, pp. 362–380; Bernhard Kuhn, "A Chain of Marvels": Botany and Autobiography in Rousseau," European Romantic Review, Vol. 17, number 1/Jan. 2006:1–20. Marc Oliver, "Lessons for the Four-Year-Old Botanist: Rousseau's 'Forgotten Science' of Childhood." *French Literary Studies.* Vol. XXX, 2004: 161–169, fails to note that the ways of the child are Rousseau's ways as philosopher and he assumes without evidence that Rousseau's science is derived from Locke. It is not so derived or it would not be called "forgotten." On botany as a means to self-knowledge see, Michael Davis, *The Autobiography of Philosophy*: *Rousseau's Reveries of the Solitary Walker.* Rowman and Littlefield. 1999.

5. Kelly, *The Collected Works of Rousseau*, 93, 97, 99, 109–109, 115, 124, 128, 130, 135, 137, 139, 140, 156, 179, 215, 215, 261.

6. Kelly, *The Collected Works of Rousseau*, 93, 109–10. Evidently, if substance means in-itself, and in-itself means unity, then, of course, substance must be indivisible. If substance is indivisible, then, of course, it has no parts. If so, then it is worthy of wonder how substance could either cause itself to move into our mind or how we could partially access it? Partial knowledge, progress in knowledge at all, would be impossible. There is no in-itself; or, if it is it is irrelevant to us; or, if it is, and it is relevant, we must "know it" all at once, or not at all.

7. Kelly, *The Collected Works of Rousseau*, 128; Kelly, C., *Rousseau on Philosophy, Morality, and Religion* (Dartmouth College Press, 2007): 266. On the current state of reductionist science in relation to form or whole as cause, see Robert Sokolowski. "Formal and Material Causality in Science," *Proceedings of the American Catholic Philosophical Association*, vol. 69 (1996): 57–67 and Richard Hassing, "Wholes, Parts, and the Laws of Nature," *Nature and Systems*, vol. 6 (1984): 195–215. Above all, for the best account of heterogeneous wholes, see Paul Stern, *Knowledge and Politics in Plato's Theaetetus.* (Cambridge University Press) 2008. On the importance of the awareness of the **future—a whole life—** and hence of "death and its terrors," that it can occur at any instant, for political philosophy, and their relation to the importance of the sentiment of **present** existence, see Gourevitch, *Political Writings*, 142: 19, 143: 21, 187: 57, 235: 11, 358.

8. Kelly, *The Collected Works of Rousseau*,108–109, 112, 116, 117, 121, 128, 140, 229, 250–252.

9. Kelly, *The Collected Works of Rousseau*, 104, 107, 123, 125, 126. On human metamorphosis, see 288 n5. Rousseau does speak of plants as living "machines" twice and employs the language of mechanism several times related to each part of the whole and the whole itself. Kelly, *The Collected Works of Rousseau*, 64, 109, Gourevitch, *Political Writings,* 115: 3, 140–141. On how natural potential and mechanism may be reconcilable, see Jean DeGroot, "Dunamis and the Science of Mechanics," *Journal of the History of Philosophy*, Jan 2008, vol. 46, Issue 1, 43–67, and Thomas Nelson Winter, The Mechanical Problems in the Corpus of Aristotle," *Digital Commons* University of Nebraska, 2007. On the other hand, Rousseau may employ "weak" mechanical language to draw potential scientists into his reformed natural science. See also Sylvai Berryman "Teleology without tears: Aristotle and the role of mechanistic conceptions of organisms," *Canadian Journal of Philosophy*, 37.3, (Sept 2007): p. 351 (20).

10. Kelly, *The Collected Works of Rousseau*, 96, 99, 130, 144, 176–177, 186, 189–190, 197, 198, 202, esp., 204–205, 207, 209, 215–216, 253.

11. Kelly, *The Collected Works of Rousseau*, 124, 115, 156. Gourevitch, *Political* Writings, 167: 18. Man must seek that which is naturally good for him.

12. Kelly, *The Collected Works of Rousseau*,103, 107, 116, 122–123, 130–131, 142, 148–149, 160, 212, 218, 230, 233, 250–252.

13. Kelly, *The Collected Works of Rousseau*, 97, 153.

14. Kelly, *The Collected Works of Rousseau*, 94, 114, 130–131, 135, 230–231.

15. Kelly, *The Collected Works of Rousseau*, 59, 97, 133, 137, 149, 155, 170, 173.

16. Kelly, *The Collected Works of Rousseau*, 144, 150, 168, 170–171, 183, 251–252.

17. Kelly, *The Collected Works of Rousseau*, 132, 150, 18, 212, 242.

18. Kelly, *The Collected Works of Rousseau*, 250, 138, 155, 156.

19. Kelly, *The Collected Works of Rousseau*, 105, 155.

20. Gourevitch, *Political Writings,* 111–122.

21. Gourevitch, *Political Writings,* 131: 2–3.

22. Gourevitch, *Political Writings,* 171: 27, 150: 33, 159: 41, 133: 7, 157: 47,159: 52, 170: 27, 205: 1, 162: 18, 159: 51, 157: 47–52, 162: 5, 166: 16, 170: 25, 166: 17, 139: 13, 142: 18, 144: 2, 141: 17. On Rousseau's view of history in relation to his "austere teleology," see Leonard R. Sorenson, "Natural Inequality and Rousseau's Political Philosophy in his Discourse on Inequality," *The Western Political Quarterly* (P.R.Q.), (vol. 43, no. 4, Dec. 1990): 763–788.

23. Kelly, *The Collected Works of Rousseau*, 97–98, 192, 225, 237, 259, and Gourevitch, *Political Writings*, 189: 52.

24. Gourevitch, *Political Writings*, 203: 131, 204: 14, Kelly, *The Collected Works of Rousseau*,64: 5, Gourevitch, *Political Writings*, 97: 18–19, 102: 33, 83: 69, 82: 65, 210: 9, Kelly, *The Collected Works of Rousseau*, 48, Gourevitch, *Political Writings*, compare 26–27, "Alone...Footsteps," "First Step," and his Title *The Reveries*

of the Solitary Walker (emphasis added), 26–27, 189: n. 11: 1; 97: 18–19, 33: 7, esp. 320, 189–190. On the status of philosophizing according to Rousseau, see Leonard R. Sorenson, "Rousseau's Socratism: The Political Bearing of "On Theatrical Imitation." *Interpretation: A Journal of Political Philosophy*. (vol. 20/2, Winter, 1992–93).

25. Kelly, *The Collected Works of Rousseau,* 262–63, note 11 above.

26. Kelly, *The Collected Works of Rousseau,* 141, 262.

27. Kelly, *The Collected Works of Rousseau,* 216, 222, 224 and Gourevitch, *The Political Writings of Rousseau*, 141: 17 and Note IX.

28. Kelly, *The Collected Works of Rousseau,* 130, 144–45, 173, 244.

29. Kelly, *The Collected Works of Rousseau,* 136, 173, 250–52.

30. Kelly, *The Collected Works of Rousseau,* 155, 206, 215, 227, 250–52.

31. See note 24 above.

32. Kelly, *The Collected Works of Rousseau,* 173.

33. Kelly, *The Collected Works of Rousseau,* 146, 214–15, 224–25, 230–31, 250–52.

34. Kelly, *The Collected Works of Rousseau,* 101, 225, 230–31, 250.

35. Kelly, *The Collected Works of Rousseau,* 173.

36. Kelly, *The Collected Works of Rousseau,* 215, n. 307, 225, 250, 252, 9, 41, 45–46, 59, 61–62.

37. Kelly, *The Collected Works of Rousseau,* 108, 144, 146, 218, 252.

38. Kelly, *The Collected Works of Rousseau,* 133, 139, 140, 150, 160, 176, 205, 236, 250, Gourevitch, *Political Writings of Rousseau*, xvii, xxx..

39. Kelly, *The Collected Works of Rousseau,* 119, 208, 132–33, 208, 230.

40. Kelly, *The Collected Works of Rousseau,* 119, 128, 139, 155, 255, 260, 264. On the current state of the debate between reductionist and nonreductionist science on the question, what is life, see "What is Life*?" Cosmos and History: The Journal of Natural and Social Philosophy*, vol. 4, numbers 1–2 (2008): 1–425. On Rousseau's critique of human physicalism or man as material mechanism, see his critique of Helvetius's *On the Mind* concerning his opinions on judgment, memory, and natural inequality of minds in relation to education, Kelly, *Collected Works of Rousseau*, XII: 204–12.

41. Kelly, *The Collected Works of Rousseau,* 177, 198–99, 239, 248–49.

42. Kelly, *The Collected Works of Rousseau,* 259, 177; 142, 148–49, 150, 161, 186, 189–90, 196–97, 202, 207, 211, 212, 230, 234, 57, 296–97 n. 21, 203, 267, 325 n. 307.

43. Kelly, *The Collected Works of Rousseau,* 174, 180, 190, 204, 209, 216, 246, esp. 194, 178.

44. Gourevitch, *Political Writings of Rousseau*, 141: 17, 348 n. 6, 108: 6, 109: 8.

45. Kelly, *The Collected Works of Rousseau*, 95, 97, 101, 116, 301n. 20, 43, 58, and Gourevitch, *Political Writings*, 114:2, 48, 55, 210:8.

46. Kelly, *The Collected Works of Rousseau,* 93, Gourevitch, *Political Writings*, 100: 22, 190: top, 73: 33, 308: 13, 10: 27, 26–27: 54; See also " yoke," "slats," "chains"115:3, 141:17, 6: 8–9, 124:1, 70–71n. The rational law of the *Social Contract* is the product of Rousseau's prudence, not nature, 114–15:1–4. The rational

law requires generalization, the cognitive root of the rational law, a "gentle yoke," a mental capacity more equally distributed among men than are calculation, and especially prudence or dialectic. The foundation of political society is human equality, an opinion derivative from the common consent of mankind, not nature. This prior consent requires consent at the foundation and in the operation of political society, 124: 3, 45: 3. A task of political philosophy is to reconcile—insofar as possible—natural inequality and equality. See note 49 below. In this regard, see Leonard R. Sorenson, "Rousseau's fulfillment of the natural public law tradition and his contribution to its demise," *The European Legacy*, vol 10, issue 5, 2005: 421–37.

47. Kelly, *The Collected Works of Rousseau,* 93; Gourevitch, *Political Writings*, 94–95: 9, 92: 1, 104–106 and especially 290: n, 3. The end of the natural form of man is progress in self-knowledge. Opinions are the obstacles to self-knowledge. They can constitute illusions or hallucinations. *Nature intended* the *natural* light of the mind to dispel them, ibid. 5: 6, 6:7, 8, 26:59–60, 124:1–2, 189–90 top.

48. Gourevitch, *Political Writings*, 118: 15, XXV, 131: 3, 188: 58.

49. Kelly, *The Collected Works of Rousseau*, 156, 189; Gourevitch, *Political Writings*, 308: 13, see note 46.

50. Kelly, *The Collected Works of Rousseau*, 130, 156, 173, 411, 192, 251, 252, 57–58.

51. Given natural potential—potent-act—a grave problem emerges regarding its full knowability, see Stern, *Knowledge and Politics*, 22–23 on Aristotle's *Metaphysics,* 1045a 9–10, 1045b21, 1048a30.

Chapter 15

Contrasting Biological and Humanistic Approaches to the Evolution of Political Morality

Steven Robinson

"Man is born free, and everywhere he is in chains." These famous words from the opening lines of Rousseau's *The Social Contract* express as a paradox what is in fact a problem for most if not all varieties of ethical naturalism: how to explain the typically wide variance between the way a theory dictates that human beings ought to live and the way most human beings are observed to live, empirically and historically.[1] As Bernard Williams perceptively noted in his *Ethics and the Limits of Philosophy* (1985), any theory which aims to produce a convergence of ethical thought on the basis of ethical reality is weakened to the extent that it cannot produce a "theory of error" which "accounts generally for the tendency of people to have what, according to its principles, are wrong beliefs."[2] This is especially the case for theories of human nature.[3] In other words, any robust ethical naturalism should not only describe an ideal pattern of behavior, but also account for and predict deviations from it in appropriate circumstances. If, for example, a theory dictates that by nature all humans ought to treat each other with equal respect, it behooves the theory also to explain the widespread human tendencies towards racism, sexism, and social hierarchy.

Many such "theories of error" will no doubt be familiar to the reader from the history of moral and political thought. The classical Greek philosophers, for one instance, speculated that ethical knowledge is exceedingly esoteric, demanding rigorous philosophical training, and so is simply beyond the grasp of most people, who are therefore doomed to lives of ignorance and paternalism. The Christian doctrine of original sin, on the other hand, implies that humans are too weak to choose right reliably even when they know what it is. By contrast, historicists like Hegel and Marx saw human nature as unfolding gradually through history toward the best form of life, such that all human beings who have ever lived are but incomplete versions of a humanity that will appear in its perfection only at the "end of history." In fact it is only because Rousseau immediately follows that same famous observation of his with a frank admission that he can supply no theory of error that his formulation of the human condition becomes paradoxical: "How did this change come about? I do not know."[4]

It is an interesting problem, particularly for all those in the modern West committed to the fundamental rightness of some version or other of a democratic and egalitarian political morality: overall, humans have been a notoriously undemocratic species. How might we account for this failure to live up to our true natures? The difficulty of wedding any kind of Platonic theory of error to democracy is well-known, since many of our most basic democratic rights exist only to protect us against incursions by elites (including religious elites) who like Greek philosophers would claim to understand the public interest better than the majority. Moreover, few in the West still cling to the old Enlightenment optimism that saw democracy as the natural outcome of a progressive universal history of mankind. And while Christian sects have variously embraced anti-authoritarian ideals and practices, those developments have been more in parallel to the emergence of modern democracy than a pattern for it. These old familiar sorts of theories of error are of little use in defending and justifying democracy today, and so a new one appears to be needed. Even the most sanguine of democrats will admit that democracy has been and continues to be a very rare and unstable historical phenomenon. In ancient Greece it lasted just one hundred and seventy years, confined to a tiny region of southern Europe. Modern democracy has exceeded that record, but so far not by much; only a few other minor examples may be identified scattered here and there through history. At no point (not even today) has a majority of living human beings ever coexisted under a range of democratic regimes—not even close. Why is that? If democratic values are somehow "natural" or "inevitable," then why are stable, functioning democracies so rare? If we are all born free, why do so many still live out their lives in chains?

There are many ways one could approach answering this question; indeed, it is such an important question that there is probably no discipline in all of the humanities and social sciences that has not tried to answer it one way or another over the past hundreds or thousands of years. All of those traditional scholarly approaches, however, have recently been challenged by a novel development within the life sciences. Ever since the publication of Wilson's *Sociobiology: The New Synthesis* in 1975, and despite setbacks in the 1980s, when human sociobiology was roundly attacked as reactionary and simplistic, the idea that all human behavior might ultimately be explained and predicted by our biology has been gaining adherents.[5] While much of this research remains novel and controversial, some of its results are undeniably interesting, especially when supported by the recent advances in genetics and genomics which have opened a new window onto human evolution.[6] Even where strong positive claims are not made for genetic determinism, proponents have begun to assert that neo-Darwinism has superseded and even invalidated previously dominant traditions within value theory such as liberalism, Marxism, and Christian humanism.[7]

There is, of course, nothing new in the proposition that some human behavior can be explained as a heritable adaptation made universal through eons of natural selection; this is easily captured by the notion of "instinct," a familiar category of behavior that is common to humans and other animal species. What was new in the 1970s and '80s was the discovery that certain types of *self-sacrificial* (i.e., "altruistic") behavior in other species could be explained entirely and convincingly as a genetically hard-wired predisposition.[8] This discovery invalidated the widely held assumption that any genetically programmed instinct must be straightforwardly self-serving (i.e., serving to further the survival and reproduction of the individual organism whose instinct it is). The urge to assist another in full knowledge that it comes at the expense even of one's own life could now be understood as an adaptive instinct and a product of natural selection. Thus was opened up for the first time the radical possibility that a wide range of behaviors might be reducible in similar fashion to genetically programmed predispositions, including those uniquely human behaviors such as religion, ethics, and politics that had always been considered antithetical to "instinct" precisely because they transcended self-interest.

Neo-Darwinists seized upon these radical implications immediately and have been pushing them hard ever since, from early forays into the evolution of ethics, through differential reproductive strategies for men and women, to the recent onslaught of anti-religious vitriol from the likes of Richard Dawkins and Christopher Hitchens. Of course, early varieties of social Darwinism, even in Darwin's own lifetime, long promoted a "dog-eat-dog" conception of selfish individualism which viewed egalitarianism as a form of misguided self-sacrifice because it awards equal power and respect to persons of unequal merit. Egalitarianism was seen to be a kind of well-intentioned degeneracy that would inevitably be weeded out by natural selection via life-and-death social competition, if not at the level of social class then at the national or "racial" level.[9] The ethical paragon of this social Darwinism was the talented egoist who shows up in various modern social theories (i.e., the rational maximizer) but who has never been a good match for the typical empirical human being.[10] However, this is not what the neo-Darwinists have in mind. Their recent elaboration of new theories of "kin selection" and "reciprocal altruism," together with a renewed focus on "group selection" since the 1980s, have striven to explain individual human self-sacrifice and genuine forms of community as kinds of heritable adaptive instinct. This has allowed neo-Darwinism to repudiate the errors of its social-Darwinist past and venture once again boldly into the field of political theory under various new rubrics such as sociobiology, evolutionary psychology, and even "biopolitics."

But has the ghost of social Darwinism really been exorcized? A consensus is now beginning to emerge that none of these recent models of selection is quite sufficient to explain democracy biologically.[11] "Kin selection" (the factor famously operative amongst social insects) cannot produce habitual altruism towards beneficiaries outside the circle of one's closest relatives: were it to do so, benefactors would inevitably lose out in competition to more selfish individuals. In other words, generalized altruism cannot emerge this way. Still, in all likelihood kin selection is responsible for those instinctive familial sentiments (including nepotism) that were noted even by Darwin in humans.[12] "Reciprocal altruism" is a factor that can be shown to produce reliable social and ethical behaviors between unrelated individuals, such as "tit for tat"—a willingness to reciprocate without cheating even when one might cheat and get away with it. This may well account for certain universal features of market-exchange behavior amongst human beings.[13] Nonetheless, repeated attempts at modeling reciprocal altruism have proved that it operates only when agents believe that "quid pro quo" is likely to be respected for each and every individual kindness. As a result, this factor can produce only behavior that is cooperatively selfish—not the sort of unselfish behavior that characterizes genuinely ethical and political action.[14] Finally, "group selection" occurs when individual instincts are so determinative of group survival that they overwhelm the effects of individual intra group competition: though they do not promote the survival of individuals or of their immediate relatives, they do preserve the group; the group's survivors then carry forward the shared "unselfish" genes into subsequent generations. However, it is now accepted that group selection is so weak vis-à-vis individual selection as to be operative only in extreme cases, such as self-sacrifice during genocidal intergroup combat. In most other social circumstances, individual selection will dominate and selfish individuals will out-compete "ethical" ones, eliminating ethical instincts from the gene pool.[15]

In the face of these sorts of results, the question facing many in the biopolitics camp now is not: How has democracy evolved, but rather how can democratic values even persist as an ideology when they demonstrably cannot be the product of natural selection? According to biopolitical theorists Somit and Peterson, these more "peaceful" sentiments in humans are products of a uniquely human cultural development, which they call "indoctrinability." This tendency derives from evolved religious behaviors whereby communities collectively cultivate fictitious and creative mythological belief systems as a means of maintaining group cohesion and identity.[16] The argument is that religious belief systems are like artistic representations of an invisible world which cannot be challenged or refuted by experience. While the value of religious behavior overall has been adaptive via group selection, it has also produced as a byproduct beliefs that can culturally evolve in opposition to other of our genetic dispositions.[17] The cultivation of nonnatural value-beliefs such as egalitarianism is therefore an evolutionary "mistake," because those beliefs will typically be contradicted by actual human behavior; but the fact that human minds have evolved to be "indoctrinable" means that we go on indoctrinating ourselves culturally to believe in them anyway. The upshot is that while we might deceive ourselves about our motives and our intentions, our evolved disposition to behave in more normal primate fashion will still usually trump our dogmas, just as is the case with most other evolved instincts. Political and moral philosophy are hereby reduced to varieties of self-deceptive mind games—and the belief in human political progress, a pure fantasy. This is predictably leading once again towards the old social-Darwinist evaluation of democracy as a kind of error, a temporary maladaptation within human biology or, even worse, a mere epiphenomenon. Somit and Peterson actually conclude the following:

> Whether we like it or not, not only our philosophers but most of humankind, throughout most of history, have taken it for granted that the few should rule. And, we suggest, despite the lip service presently given to a near vacuous "democracy," they probably continue to do so.[18]

Are we perhaps standing on the threshold of a social-Darwinist resurgence?

THE SCOPE OF THE PROBLEM

It is worth noting that moral and political theories, regardless of their culture of origin, have always represented human beings as psychologically complex, in that we are torn by competing tendencies toward both selfishness and unselfishness. Pessimistic theorists portray us as "naturally" selfish beings who sometimes appear (or pretend) to behave unselfishly; optimistic theorists say that we are "naturally" unselfish, but that we are often so compromised or perverted as to behave selfishly. More astute empirical observers acknowledge that a mixture of the two is typical and that both extremes are exceedingly rare, and then they add in various accounts of the relative importance of nature versus nurture. If new sociobiological theories are to contribute anything to these more traditional discourses, then at the very least we ought to expect of them that they reflect and explain this moral and psychological complexity. Indeed, if human altruism were understood to be as instinctive as that of the social insects, we might wonder why we have such a thing as normative discourse at all. Ants and bees may very well be genetically programmed to sacrifice themselves altruistically on behalf of their kin, but they show no signs of having any qualms about doing so. Humans, by contrast, have qualms about all sorts of moral imperatives. And yet if the ultimate cause is understood to be the same in both cases, and as having been equally determinant of these species' past survival, then why is it that humans are so conflicted? Why is it that we even recognize and set apart certain values and imperatives as "ethical" instead of simply acting them out unselfconsciously (as the ants appear to do)?

In all likelihood human history could deliver up real examples of just about every conceivable kind of human political relationship, however brief or experimental. Only some of those have been subjected to sophisticated theoretical analysis, however, and many of those became possible only with the advent of the radical individualism that marks the birth of modernity. Nonetheless, the scope of the problem may be neatly delineated by means of three early modern theorists who between them exemplify the extremes and the middle ground of the debate over human ethical nature in the context of political morality. Niccolò Machiavelli (1469–1527) represents humans as all perniciously selfish, and as struggling competitively for power and control over each other at all times; he advocates absolute despotic rule by a single dominant egoist. Etienne de la Boétie (1530–1563), an anarchist, argues just the opposite: that humans are all so nearly equal that common sense demands none be given pre-eminence over another, except insofar as his talents may be employed for the service of his fellow men. Finally, William Shakespeare (1564–1616), in his play *Measure for Measure,* dramatically lays out a rationale for the exercise of state power by outstanding individuals within ethical constitutional constraints. These three works, by authors whose lifetimes, laid end-to-end, span the birth of modernity, still define the scope of our present-day political imagination, and hence of the field of human political reality within which neo-Darwinism too must operate. They can effectively capture and frame the present-day debate concerning authoritarian versus democratic biological instincts.

The Prince, written in 1513, is Machiavelli's guide to success in politics, composed to gain favor with the new prince of Florence. It describes contemporary Italian political life as a desperate, high-stakes competition for security and power, and in the process it articulates a deeply cynical but profoundly influential theory of human nature. Far from being the immoralist he is often described as, Machiavelli unfolds in detail the prescriptions of an ethical egoism designed to appeal to the privileged, powerful "alpha males" of his day. Accordingly, all humans are selfish and seek power for themselves, and so are false and unreliable to others. However, there is a wide spectrum of abilities in people, and those with strength and talent can and should exploit these differences to their advantage. Power is sought primarily as the means to personal security, but then also to whatever end one might wish to define as "wellbeing," such as possessing wealth and property. Only fear renders people reliable, and so *The Prince* is largely a textbook on how to employ intimidation, deception, and violence to gain the submission of those around you, and to isolate and eliminate potential competitors for dominance:

> Let a prince therefore aim at conquering and maintaining the state, and the means will always be judged honourable and praised by every one, for the vulgar is always taken by appearances and the issue of the event; and the world consists only of the vulgar, and the few who are not vulgar are isolated when the many have a rallying point in the prince.[19]

The result is a virtue-ethics which prizes individual success by any and all means, where success in every case means the submission of one's entire community to one's dominance and control, through fear. By winning this submission, the prince proves his right to dominate. When Machiavelli advises that the prince must "learn how not to be good," he refers to not to his own sense of "good" (the successful prince is "good" by definition), but to the popular morality which is wholly invalid because it is based on false and unrealistic conceptions of human nature.

While nowadays it is empirically difficult to accept Machiavelli's blanket condemnation of all humanity as selfish and mean, it is undeniable that he has accurately described the behavior of many political leaders throughout history; indeed, his historical evidence is striking for the sheer magnitude of cases he can point to. History since then has supplied many more. Moreover, his textbook for political success has stood the test of time: it is widely respected as a guide to methods that actually *work* in the real world. It is easy to read Machiavelli's account of human nature in Darwinian terms. It is none other than the most pessimistic version of social Darwinism, whereby each individual is engaged in a struggle to survive and thrive in what Hobbes would later call a "war of all against all."[20] Like a latter-day Thrasymachus, Machiavelli's claim is that the fictitious theories of egalitarian political morality which humans use to conceal this truth are themselves just a cynical attempt by the weak to turn tables and disempower the strong. For the sake of manipulating the many, the true prince must appear to respect this popular morality, but pay it lip service only. While present-day neo-Darwinists can easily escape the extreme individualism of Machiavelli through a more sophisticated account of local morality as a product of group-selection, it is not clear that they can avoid naturalizing atavistic intergroup conflict in a Machiavellian vein.

La Boétie, born just a few years after Machiavelli's death, delivers a striking rebuttal and critique of Machiavelli's theory of human nature. He begins by rejecting the notion that people are much different from each other. Yes, some may by birth be stronger, richer, smarter, and better looking than others, but none of these things is so impressive as to justify the former being given the power of life and death over the latter. La Boétie points out that the reason princes must rely on deception is that they can never actually bully and intimidate the masses, but only individuals. To control the masses, the prince must rely on the willingness of the majority to accept his legitimate right to rule them. Never, though, argues La Boétie, has there been a prince with any such legitimate right to rule. For each of Machiavelli's historical cases, La Boétie can counterpose the case of a prince who horribly abused his people. The problem he sets himself in his aptly titled *Discourse of Voluntary Servitude* (1550), is to understand why human beings so uniformly submit themselves to the rule of rapacious princes. His answer is that civic freedom requires the widespread development of critical reason, together with the opportunity for the masses to communicate with each other. The princes understand that to control the masses they need only supply and censor public media of a sufficiently diverting sort that the development of critical reason is stultified, and the masses are thereby rendered perpetual and obedient minors.

To this analysis La Boétie adds a perceptive sociology of power, whereby a small but perverse fraction of the wider populace—namely, those who seek power over others—are drawn together into a hierarchy of dominance and submission, leading up in a pyramid to the sole figure of the prince. Each level within this pyramid places itself at the mercy of the level above in exchange for the right to abuse the levels below, and in the hope of advancing even higher through "good service." Thus does the prince assure himself of a small army of slaves sufficient to seek out and destroy any amongst the masses who can think for themselves individually and object to his power. In such conditions there is little hope for freedom to develop and political reform to follow, with the result that human nature is collectively stunted and frustrated. La Boétie's goal is to activate the seed of reason within us all sufficiently to stimulate intercommunication of the masses leading to the collective throwing-off of domination by others:

> From all of these indignities, such as the very beasts of the field would not endure, you can deliver yourselves if you try, not by taking action, but merely by willing to be free. Resolve to serve no more, and you are at once freed. I do not ask that you place hands upon the tyrant to topple him over, but simply that you support him no longer; then you will behold him, like a great Colossus whose pedestal has been pulled away, fall of his own weight and break into pieces.[21]

According to La Boétie it is reason that makes us free, but reason must be active within the collective if it is to lead to action and not just individual despondency. Like children growing up and out of the tutelage of their parents, all people must mature into their freedom from rule by others, but they need each other to do it. The power of princes is based on preventing that intercommunication and maturation of their peoples.

La Boétie is not so simple as Machiavelli to read in Darwinian terms, but he has the advantage of complexifying the view of human nature: some seek peace and equality, others seek dominance and hierarchy. While La Boétie clearly indicates that the development of reason is common to all human beings, and that we are thereby all by nature free and equal as citizens, he fails to diagnose the cause of some people's desire to join the rapacious power-pyramid. By contrast with Machiavelli (who would say that the power-pyramid contains everyone but that most remain forever at the bottom), La Boétie believes it is some sort of unnatural pathology that leads a small fraction of the people to power-seeking, while the vast majority want nothing more than to live and work in peace. From a Darwinian perspective, this division of the population into power-seeking and non–power-seeking factions would seem less likely than a common principle as an outcome of natural selection; however, we shall return to these two options more closely below.

Our third author, Shakespeare, born just the year after La Boétie died, shares La Boétie's recognition that people divide into opposed characters—though for Shakespeare in *Measure for Measure* the characters in question are those of "goats" and "sheep" (sinners and the redeemed). The title of the play refers to Matthew 7:1–2 ("Judge not that ye be not judged . . ."), and indicates the play's deep concern with issues of punishment, forgiveness, and legal authority. We are presented with three persons who seem paragons of virtue. One, the Duke, puts the other two to the test: of those two, one (Angelo) falls by temptation into grievous sin while the other (Isabel) remains pure. The rest of the cast presents a variety of sinners, from fairly innocent through to pernicious. Angelo sins when deputized by the Duke with his full power, while the Duke supposedly absents himself on a trip. In reality the

Duke stays in disguise and observes Angelo abuse his power. While Angelo's crime (against Isabel) is severe, he is nonetheless forgiven by the Duke at the end, along with most of the guilty others; a few, however, (the unrepentant pimps and prostitutes brought repeatedly before justice) are not forgiven, and are sent instead to prison. One clear message of the play is that punishment is to be used only rarely and in those cases where persons repeatedly demonstrate their unwillingness to do what the law demands; forgiveness is for those who commit crimes through weakness, but who fundamentally accept the legitimacy of the law and can be rehabilitated through public exposure for their crime.

Shakespeare thus indicates, like La Boétie and contrary to Machiavelli, that a prince cannot control his people through fear alone, but must count on the people's desire to obey. Unlike La Boétie, however, Shakespeare contrasts the virtuous prince (the Duke) to the selfish prince (Angelo) and shows how the people's love for the former can be justified. In a paraphrase of Matthew, the Duke spells out the difference:

> He who the sword of heaven will bear
> Must be as holy as severe;
> Pattern in himself to know
> Grace to stand and virtue go;
> More nor less to others paying
> Than by self-offenses weighing.
> . . .
> O, what may man within him hide,
> Though angel on the outward side![22]

The upshot is that society depends utterly upon the basic willingness of people to do what is right and resist the temptation to do wrong; but because people are weak, they sometimes give in to temptation and must be corrected—by the state. The prince (i.e., the state) is required publicly to set the standards people are expected to meet, and to sit in judgement of them on that score. However, punishment is a last resort only, and corresponds to Machiavelli's instrumentality of fear. It is to be used only on those who have demonstrated their lack of "love" for the law, and who can be controlled only by being threatened. The power that can legitimately make and enforce the law is neither absolute nor despotic: it must itself be "holy" and hence deserving of the people's obedience. That is, it must be constitutional. Angelo, by abusing his princely power, violated the implicit moral constitution that lies behind all legitimate state power.

Of special note here is the fact that, for Shakespeare, authoritative institutions are required to embody the moral vision of propriety shared by all of the people. It is not enough to leave people to their own devices in behaving properly themselves and in correcting each other. The play shows how the people, if left to themselves, will grow ever more licentious until "Liberty plucks justice by the nose;/The baby beats the nurse, and quite athwart/Goes all decorum."[23] Accordingly, we may interpret Machiavelli's prince, who relies entirely on fear to control his people as corresponding to Angelo, the unjust prince who rules without legitimacy in Shakespeare. Likewise, we may interpret La Boétie's "power-pyramid" as populated by those persistent sinners from Shakespeare who do not love the law and who themselves need to be controlled by threats and, if necessary, by righteous violence. Shakespeare's play suggests that the problem with La Boétie is that he has never seen a good prince, and therefore believes that all princes are enemies of the people. Shakespeare argues that civil society depends utterly on two factors: that just, constitutional governments exist, and that the people can, for the most part, be relied upon by nature to obey them willingly, and not by coercion. Of our three authors, it is Shakespeare who is the most complex, as he appears to naturalize both authority and hierarchy, on the one hand, and popular moral self-restraint on the other. Unjust governments (and any government recommended by Machiavelli), occur when those in positions of authority do not themselves exhibit effective moral self-restraint. It remains an open (and very sensitive) question whether Shakespeare himself might have advocated action by the people to remove an unjust prince from power, or whether he functioned instead as La Boétie might have described him: i.e., as a mere apologist for unjust power and part of the prince's machinery for hoodwinking the people (after all, his troop was named The King's Men).

A PROPOSED DARWINIAN SOLUTION

This very debate, I shall argue, can be taken as reflecting the competition between the factors of cultural and rational choice on the one hand, and genetic evolution on the other. The crux of my argument, however, comes from insights in comparative primatology and cultural anthropology. In short: the source of human moral complexity is likely prehistoric, but traditional humanistic theory has restricted itself to historical exemplars (as illustrated by Machiavelli and La Boétie) in its striving to understand human sociality. While strictly neo-Darwinian approaches may be coming up short in their attempts to reflect and explain the complexity of human moral and political psychology, that may be because of their reliance on mathematical models of natural selection. By drawing upon recent studies of nonhuman primate "politics," we can discern what are potentially direct analogs with familiar aspects of traditional humanistic political theories. This raises the possibility of explaining the evolution of political morality without either reducing it to genetically determined causes or committing the naturalistic fallacy.

While the mostly speculative tradition of humanist political scholarship has ignored human biology and based itself empirically on the comparatively small sample that is provided by human "history," proponents of "biopolitics" point instead to the deep prehistory of humanity which stretches almost 200,000 years into the past and blends continuously into the genetic evolution of other species, including the other primates and now-extinct hominids. What was human life like during that long prehistory? There are no written records to tell us and only scant archaeological evidence to go by, insufficient to reconstruct social structure or behavior in any detail.[24] But a great deal of evidence does still exist: not only genetics, but comparative primatology and the record of physical anthropology. This evidence points to a very different account of humans in the "state of nature" than what we typically find in humanistic theories.

All nonhuman apes are hierarchical animals whose social behavior is governed by relations of dominance and submission, though to a degree that varies by species. They are territorial animals who compete violently with other groups of their own species for resources. Those most closely related to humans, the chimpanzees, are the most despotic of all: every adult male enjoys a status higher than all females, and both males and females compete with others of the same sex for positions within a linear hierarchy, sometimes violently. Those at the top of the hierarchy enjoy considerable advantages in access to food and mates; their genes are disproportionately reproduced in all subsequent generations, and to a degree that corresponds to their rank;[25] they continuously dominate, control, hector, intimidate, ostracize and occasionally even kill those who challenge them for status and those who submit to them alike. A wide range of bodily characteristics (e.g., size and shape of teeth, thickness of skull, sexual dimorphism) are identifiable as the physical correlates of these social relations, and these features may be read in the fossil remains of early humans and others just as they are in the kinetics of living apes. Without a doubt, primate nature has been honed through millions of years of intense natural selection for success in intraspecific dominance, aggression, and intimidation, and this is the context out of which early humans first emerged.[26]

To judge by our physical traits alone, humans ought to fit neatly into this pattern of despotic social organization, likely somewhere in the mid-range of existing apes. And while human social organization in deep prehistory eludes us, there is certainly no shortage of evidence for despotic violence from recent human history, let alone the daily news. Most of that evidence appears to confirm that these primatological insights can describe a significant portion of human behavior. To some of the proponents of neo-Darwinian social theory, this evidence reveals the true reality of our nature and our inheritance, i.e., despite our protestations to the opposite, deep down we are violent, aggressive and largely selfish apes who desire above all to dominate others but who also submit to domination by others who intimidate us. This is no less than a variation on the account of human nature described by Machiavelli, and illustrated by innumerable historical examples.

Is it possible to bring these competing accounts of human nature together—the biological and the humanistic? As we have seen, political philosophy has long recognized fairly profound divisions within the human person, so much so that the human psyche has often even been said to be "at war with itself." Shakespeare (above) is a prime example of this view, but he is in good company: for Plato as for most of the Western tradition following him, this division was accepted and viewed as a struggle between reason and passion, or soul and body. The key to seeing how these classical treatments have anticipated recent discoveries in human biology is provided by evidence from *cultural* anthropology. Just as neo-Darwinian biologists look to genetic, physical, and behavioral parallels between species in order to construct theories of human political behavior during prehistory, anthropologists look to the behaviors of surviving prehistoric peoples, and extrapolate backward through time. In fact, one of the most remarkable aspects of observed prehistoric cultures is the elaborate and consistent effort that they put into maintaining an egalitarian ethic—far more so than in "historical" populations and in direct contradiction to what their primate biology alone would supposedly predict. Christopher Boehm has noted how this egalitarian ethic is virtually universal in existing prehistoric communities and in all likelihood was universal throughout human prehistory, at least since the advent of complex language.[27] He indicates that even chimpanzees have been observed to exhibit sporadic instances of egalitarian action where groups of submissive apes collectively stand up to dominant individuals, effectively turning the status and power hierarchy upside down—momentarily. Chimpanzees however, do not possess any means to extend these events beyond the moment.

Early humans, by contrast, possessed several features that pre-adapted them to the establishment of what Boehm calls stable "reverse dominance hierarchies," in which low-status individuals collectively assume ongoing political control over high-status individuals. Most notable here was the invention of lethal weapons that can kill from a distance (probably 100,000 years ago), allowing low-status individuals easily to dispose of dominant intimidating males. Both humans and chimpanzees demonstrate not only the desire to dominate others, and the willingness to submit, but an additional emotional response: resentment of submission. This resentment means that their submission is never total and dependable, and that leaders must always expect coups from below (as Machiavelli indicates). According to Boehm, however, the final element that enabled humans to develop and maintain reverse dominance hierarchies was language. Language both empowered low-status individuals to build coalitions against their more intimidating dominators and allowed them to memorize and track through narrative the dominance-seeking actions of high-status individuals. The upshot of Boehm's case is that humans are indeed genetically predisposed to selfish, violent, dominant, and intimidating primate behavior, but that low-status humans have found ways to act upon their resentment of submission through coordinated collective action. Through the medium of complex language, humans overcame their submissive primate emotion of fear by standing and working together with their low-status peers to resist oppression by intimidating individuals. This was a major cultural elaboration of a separate pre-existing minor genetic tendency—the tendency to resent submission.

In short, the development of language allowed human communities to articulate an ethic of cooperative opposition to dominance by high-status individuals. This ethic created conditions (e.g., more equal sharing of meat, more equal access to mating opportunities) whereby intergroup competition began to outweigh intragroup competition. Consequently, group selection became a much greater factor than it typically had been, and altruism-encouraging genes then began to be perpetuated.[28] To the extent that human communities with reverse-dominance hierarchies could successfully out-compete communities with standard ape dominance hierarchies, altruistic genes could become universal. The evidence that reverse-dominance hierarchies were almost universal in prehistoric human communities powerfully supports the claim that they out-competed and replaced earlier genetic hierarchies. At the same time, high status individuals did not lose their innate disposition to dominate and control: rather, they were forced to adapt themselves to the reverse-dominance conditions that might lead to their death at the hands of their own communities if they were perceived to be taking power (prehistoric communities regularly assassinate high-status individuals who are perceived to be too haughty[29]). Their status, however, which reflects their superior natural abilities, makes them too valuable for the community to dispense with them frivolously. The result is a fine balancing act where opposed selfish and altruistic genetic predispositions are both preserved within all members of the human species: humans experience psychological dissonance as they are inclined (genetically) to be both respectful of moral obligations (maintained by an articulate social ethic) and tempted by unethical self-aggrandizement.[30] Moreover, the pre-institutional means by which humans articulated and effected an ethical conscience was subversive collectivization via gossip, whisper campaigns, slander, and invective.

That is precisely how we might guess that La Boétie's anarchistic and undifferentiated free communities would maintain their freedom from would-be tyrants if ever they could depose them to begin with. On the other hand, the proper diagnosis of the power-seeking deviance in La Boétie is not as a pathology, but as the natural, selfish instinct to dominate that exists within every human's "ambivalent" heart, but which comes to the surface primarily in those talented individuals who possess the abilities required to intimidate other individuals into submission. Machiavelli on the other hand is merely the unabashed spokesman for the instinct to dominate, and he ruthlessly dismisses any tendency toward collective, moral values (including his own natural inclination: princes must "be taught" not to be good). The most interesting case of our three though is Shakespeare: here we see high-status individuals acquiring dominance by using their power *to intimidate in the service of collective, ethical values.* The good-prince/bad-prince contrast from *Measure for Measure* becomes the conflict between those high-status individuals who serve others and those who serve themselves. The good prince is a new kind of human being whose dominance has been institutionalized by the very people he dominates, and doubtless corresponds historically to the appearance of the idea, and the very possibility, of constitutional government. In such a dominant individual the same battle between opposing natural predispositions is played out psychologically as is played out by his subjects socially.

The implications of these insights for political and moral theory are quite striking. For at least 50,000 years humans lived in small-scale, territorial societies that maintained egalitarian political structures in the face of continual attempts by powerful individuals to dominate the group from within. Then, almost overnight, this successful collectivist political balance (a "reverse dominance hierarchy") was completely upset with the appearance of settled agricultural communities at the end of the last ice age (~10,000 years ago)—and only then did human "history" begin. Somehow, likely by means of the private accumulation of capital made possible by settlement, dominant individuals learned how to beat the reverse dominance hierarchy system and return human society to the more standard, linear chimpanzee-like dominance hierarchies that have been almost universal throughout the period of settled human civilization: this gives us the political sociology of La Boétie. Ironically, this comparatively recent reversal of fortune has been so complete that humans in the historical period have become by far the most despotic of all of the apes:

the collapse of the social effectiveness of the culturally maintained prehistoric egalitarian ethic has restored genetically based ape-like dominance hierarchies with a vengeance. However, the subsequent, slow, and painstaking development of anti-authoritarian and democratic values *within* institutionalized hierarchical societies (what we call democracy) can be seen as a reassertion of the prehistoric egalitarian ethic, this time working through "channels."

Which, then, may we deem to be the more "natural" option for humans: hierarchy or equality? Political philosophers have been misled somewhat by their concentration on historical evidence, because history does not reflect the fact that extreme despotism was a relatively new thing in the first cities, and that the persistence of social dominance by elite groups throughout history to this day has been maintained by much deeper and genetic ape-like means: the leaders' ability to achieve mass submission through public demonstrations of violence and intimidation. Philosophers who understand the evolution of political morality to be the inevitable march of reason from prehistoric despotism to ultra-civilized freedom may well have their story backwards—for the present at least.

ENDNOTES

1. Jean-Jacques Rousseau, *The Social Contract and Discourses* (New York: Dutton, 1973, trans. G.D.H. Cole), 1.
2. Bernard Williams, *Ethics and the Limits of Philosophy* (Cambridge, Mass.: Harvard University Press, 1985), 151.
3. Williams, *Ethics,* 152.
4. Rousseau, *Social Contract,* 1.
5. Edward O. Wilson, *Sociobiology: The New Synthesis* (Cambridge, Mass.: Belknap Press, 1975).
6. Nicholas Wade, *Before the Dawn: Recovering the Lost History of our Ancestors* (New York: Penguin Press, 2006) provides an overview of the latest advances and some of their most striking implications.
7. John Gray, *Black Mass: Apocalyptic Religion and the Death of Utopia* (Toronto: Doubleday Canada, 2007) is a prime example.
8. W.D. Hamilton, "The Genetical Evolution of Social Behaviour, I and II," *Journal of Theoretical Biology* 7 (1964): 1–52.
9. Gregory Claeys, "The Origins of Social Darwinism," *Journal of the History of Ideas* 61, no. 2 (April 2000): 233–235.
10. Critiques of this theoretical model of the individual (a model which is still central to economic theory), especially feminist critiques, have long ago shown its limitations compared to a more socially and physically "embodied" conception of the human agent. But it may still be a perfect match for that chief institutional product of the social-Darwinist era, the corporation, which see:

Joel Bakan, *The Corporation: The Pathological Pursuit of Profit and Power* (Toronto: Penguin Canada, 2004).

11. Christopher Boehm, *Hierarchy in the Forest* (Cambridge Mass.: Harvard University Press, 1999) 197–207.
12. Boehm, *Hierarchy,* 220.
13. This tendency noted and theorized at length in: Jane Jacobs, *Systems of Survival* (New York: Vintage Books, 1992), esp. 33–50. See also, Paul Seabright, *The Company of Strangers* (Princeton, N.J.: Princeton University Press, 2004).
14. Boehm, *Hierarchy,* 202.
15. Boehm, *Hierarchy,* 204–219.
16. Albert Somit and Steven A. Peterson, *Darwinism, Dominance, and Democracy: The Biological Bases of Authoritarianism* (Westport, Conn.: Praeger, 1997), 77–84.
17. In particular, see Somit and Peterson, *Darwinsim,* 93: "The present enthusiasm of democracy—or for any other belief—can be viewed as another example of Homo sapiens' unique capacity for self-indoctrination—that is, the ability to create values and then act in pursuit of them, even when those values are inconsistent with or actually opposed to our genetically transmitted behavioural tendencies."
18. Somit & Peterson, *Darwinism,* 94.
19. Niccolò Machiavelli, *The Prince* (New York: New American Library, 1952, trans. Christian Gauss), 94.
20. In an eerie confirmation of the reproductive success of Machiavellian princedom, geneticists have been able to show that today a whopping eight percent of all males throughout the lands of the former Mongol empire (i.e., 0.5% of the total world population) carry the Y chromosome of Ghengis Khan: that is, they are his direct descendants. The world's second-most common Y chromosome would appear to be that of Giocangga, the patriarch of the Manchu rulers of the Qing dynasty: Wade, *Before the Dawn,* 235–237.
21. Etienne de la Boétie, *The Politics of Obedience: The Discourse of Voluntary Servitude* (Montréal: Black Rose Books, 1997, trans. Harry Kurz), 52–53.
22. William Shakespeare, *Measure for Measure* (New York: Signet Books, 1963), III, ii 264-274.
23. Shakespeare, *Measure,* (I, iii 29–31).
24. Chris Gosden, *Prehistory: A Very Short Introduction* (Oxford: Oxford University Press, 2003) 18–30.
25. See note 20, above, for a human comparison.
26. Boehm, *Hierarchy,* 125–148.
27. Boehm, *Hierarchy,* 90–124 and 187–91.
28. Boehm, *Hierarchy,* 205.
29. Boehm, *Hierarchy,* 121.
30. Beohm refers to this phenomenon as "innately structured ambivalences," in *Hierarchy,* 225–47.

Dialogue of the Sciences and the Humanities

Ann Ward

The history of philosophy is only very rarely understood in connection with similar developments in the history of the sciences.[1] Yet, as the contributions to this volume have shown, the connection is crucial as philosophy and science did not develop in mutual isolation from and indifference toward each other. Rather, developments in philosophy and developments in the sciences partake in a complex interaction that mutually impacts both. Moreover, not only do scientists and philosophers influence each other's ways of thinking, but also the same thinker often engages in both fields of study over the course of his or her career or even in a single work. As the chapters in this volume have demonstrated, philosophers' scientific ideas often shape their ethical and political ideas, and their understanding of the ethical and political world of human beings often shapes their understanding of the natural, material world.

The interaction between political and scientific ways of thinking is illustrated by Al-Maini in his analysis of Presocratic hylomorphism. Al-Maini argues that the natural philosophers, shaped by the Greek polis in their thinking, projected an image of the polis onto nature. As the polis provided a home for citizenship and political competition, so matter provided a home to forms. The polis, in turn, through the lens of Presocratic natural science, became the instantiation in the social world of nature's character. In my discussion of Plato's *Phaedo*, I argue that for Socrates, ethical and political questions such as whether or not the human soul is immortal will inevitably turn into questions of natural science, such as how and why the cosmos came into being. Moreover, the failure of Presocratic natural science to reveal to Socrates the ultimate source of causality in the cosmos leads to his articulation of the Ideas or Forms. The Ideas hold out the possibility not only of a stable universe, but also of a discursive form of politics within the polis that would allow for human flourishing. Zoller argues that for Plato, not only does mathematical-scientific understanding improve our knowledge of the cosmos but his "science of living well" also improves our knowledge of ourselves in relation to others. Articulated in the *Republic* and the *Gorgias,* Plato's science of living well requires a physical and psychic training that teaches citizens to master the art of measurement, putting an end to political conflict and war with the rest of the world's citizens.

In chapter 4, Younesie analyzes Aristotle's *Politics* in light of his *Physics.* According to Younesie, Aristotle's understanding of politics, especially the transformation of human beings into citizens through the agents of the constitution and the education of the polis, is deeply influenced by his understanding of natural science and its two basic concepts matter and form.

The chapters in part II also explore the complex dialectic between natural science and the ethical and political concerns of philosophy. Weinman analyzes Ptolemy's claim that there is a causal connection between the mathematical study of the heavens and the ethical pursuit of a well-ordered soul. For Weinman, Ptolemy's mathematical account of the uniform if anomalous motion of the sun does not reveal physical movement but the very being of order, the grasping of which contributes to the ordering of our souls. Meguid argues that for Averroës, the science of psychology provides the foundation for judging what is demonstrable and thus comprehended in the science of politics, and what is undemonstrable and thus must be excluded. Moreover, his search for a scientific basis of politics, according to Meguid, led Averroës to articulate a liberal and individualistic conception of the human intellect.

The complex interaction between developments in natural science and developments in philosophy is further explored in part III. Schaefer argues that Montaigne espoused a deterministic and materialistic conception of nature that came to underlie a progressive natural science and particularly a modern science of medicine in which the only ills to be avoided are physical pain, disease, and premature death. Coinciding with his support of this new science was Montaigne's advocacy of a popular public skepticism toward religious opinions and a more refined philosophic skepticism that endorsed a libertine way of life. Raymond demonstrates that Descartes was also highly involved with the new science of medicine, engaging with Harvey in a debate over the motion and hardening of the heart. Yet, Raymond shows that Descartes' explanation of the heart's motion results from his unique development of "premodern" logical intuitions drawn from Parmenides' method, Aristotle's logic, and Euclid's *Elements.*

Part III also includes four chapters that address the philosophy of Thomas Hobbes. Both Ulrich and Bradshaw argue that Hobbes sought to illustrate that he was correct about the "natural condition" of human beings by explicitly making proofs based on a natural science that posited the materialistic, mechanistic workings of the human mind. From this epistemological basis Hobbes, Bradshaw argues, tried to formulate a ground for politics that was scientific and universally true for all people in all time. Hobbes' "one world" politics, according to Bradshaw, thus rests on a world science that is based on a questionable foundation and actually makes Hobbes, unlike Aristotle with his philosophy of purpose, a surprisingly *unpolitical* thinker. For Lemetti, the core of the disagreement between Hobbes and Aristotle is Hobbes' rejection of Aristotle's essentialism. Unlike Aristotle, who posits two substances in nature, matter and form, Hobbes posits only one, matter. Lemetti argues that the critique of Aristotelian essentialism was in the service of Hobbes' attempt to articulate a more pluralistic view of human nature and its purposes. Makus, like Bradshaw, argues that Hobbes' scientific ground for politics is questionable. Attempting to impose the artificial, covenantal basis of political society on to his account of biological family relations, Hobbes cannot adequately explain the perpetuation of the human species and the reproduction of the commonwealth.

In part IV, Ward analyzes the relationship between Locke's natural philosophy and his political philosophy. Locke's critique of the Aristotelian notion of substance and his narrowing of the range of what can be considered knowledge means, Ward argues, that for Locke both natural and moral philosophy operate in the realm of probability. The probabilistic nature of both natural and moral philosophy in Locke's account has implications for liberal society and politics. Locke, Ward concludes, leaves liberal society with a "divided mind," giving rise to the possibility of factions moral and religious doing battle with factions scientific and secular over the very meaning of evidence, knowledge and proof. Sorenson argues that a careful exploration of Rousseau's botanical works reveals that he, in contrast to Locke, rejects modern natural science and returns to the philosophy of purpose, or "teleology," of the ancients. Thus, unlike reductionist modern science, Rousseau argues that plants have structured motion toward certain ends: self-reproduction and death. Moreover, for Sorenson the language of natural potentials in Rousseau's botanical works has its analogue in Rousseau's moral and political works. In his *Discourse on Inequality Among Men*, Rousseau argues that the human psyche has the capacity for internal self-perfection and external perfectibility, self-knowledge coming to light as the peak of human possibilities. Sorenson concludes that for Rousseau it is in the study of botany that the emergence of the highest purpose of political philosophy—the quest for knowledge—first appears. In the final chapter of the volume, Robinson reads three early modern political thinkers—Machiavelli, La Boetie, and Shakespeare—in light of the explanation of human behavior espoused by traditional and contemporary evolutionary biology.

Many of the chapters in this volume also move beyond the relationship between science and philosophy to problematize the received view of scientific objectivity. According to the received view, objectivity is achieved when the scientist or thinker is abstracted from the process of scientific discovery and justification. Objectivity thus entails that all social, political, religious and personal values and preferences be excised, or that the scientist remain "neutral," as it were. This ensures that data is collected impartially and expertly, and that there are logical and rational connections between data and the scientific theories drawn from it. Thus, according to the received view, when we say that a truth or conclusion is "objective," we are asserting that the empirical evidence, or the "facts," and not the scientist's subjective personal or cultural commitments, support the conclusions drawn.[2]

W.V.O. Quine famously critiques the received view of scientific objectivity. He argues that how sensory data is selected and organized into "facts" that are relevant to a given hypothesis or problem in contrast to facts that are not, is determined by prior beliefs that condition what scientists are willing to accept about reality. For Quine, therefore, scientific epistemology is influenced by a "scientific heritage" that may be rooted in beliefs that are not scientific at all.[3] Thomas Kuhn, like Quine, explores the ways in which nonscientific values are inherent in science, and argues that what he calls "paradigm" acceptance and the practice of science do not rest solely on empirical evidence.[4] Rather, Kuhn maintains that the progress of science has been strongly influenced by personal or cultural commitments that are not scientific. To illustrate his thesis, Kuhn points to the nearly two millennia prior to the "Copernican Revolution" when both the ancient Greeks and later the Catholic Church held to the "Ptolemaic" or geocentric model of the universe.[5] Loyalty to the geocentric system, according to Kuhn, was not based on empirical evidence. Rather, in the case of the Greeks, it reflected the philosophic commitment to an orderly cosmos of circular motion and uniform speed accessible to reason. For the Catholic Church it represented the theological commitment that humanity was God's highest creation and thus duly placed in the center of His universe.[6]

Although not explicitly addressing contemporary philosophy of science and scientific objectivity, chapters in this volume, like Quine and Kuhn, implicitly call into question the abstraction of the philosopher from the acquisition and articulation of scientific knowledge. Rather, the impact of the ethical and political commitments of philosophers in the history of philosophy on their scientific ideas concerning the natural, material world is openly explored. For instance, as we have seen, the political commitment to the polis as the site of a discursive politics that could negotiate a plurality of political interests led the Presocratics to develop a natural science of matter and form, Socrates to articulate his theory of Ideas, and Plato to develop his "science of living well." Moreover, Aristotle's commitment to the polis as the highest organizing principle of human life was reflected not only in a hylomorphism begun by the Presocratics, but in his teleological natural science or philosophy of ends or purposes. The ethical commitment to the psychic harmony of a well-ordered soul, an underlying motivation neglected by Kuhn, led Ptolemy to espouse the mathematical study of the cosmos that pointed to the uniform if anomalous motion of the sun as the very being of order.

When the volume turns to the early modern and Enlightenment articulation of a new science and a new politics, the authors, either explicitly or implicitly, suggest that Montaigne, Descartes, Hobbes and Locke adopted a materialistic, mechanistic understanding of the natural world. Denying the existence of nonmaterial substances in nature, these early modern and Enlightenment thinkers rejected the teleological or purposive understanding of nature of the ancients and medieval thinkers. Bottici, as we have seen, claims that this thoroughgoing materialism and nonteleological concept of reality espoused by the Enlightenment was motivated by the attempt to liberate the new science from the authority of a Christian Aristotelianism that also suppressed the human autonomy that would ground a new liberal politics.

Is Bottici correct? Does the materialism and mechanism of the early modern and Enlightenment philosophers derive from their prior prescientific, anti-theological and proliberal moral and political commitments? After all, it appears that Rousseau's rejection of modern natural science parallels his critique of modern liberal politics. Thus, although, as our authors suggest, Montaigne, Descartes, Hobbes and Locke are careful to maintain the distinction between rational truth and myth, is this rational truth "objective," or does it embed the subjective moral and political preferences of the philosophers who articulate it? The chapters in this volume suggest that further research into this question would be both fruitful and necessary. For instance, Robinson suggests that the attempt to justify their commitment to democratic values and an egalitarian political morality has guided if not determined much of the research of contemporary neo-Darwinians and cultural anthropologists. This suggestion does not seek to undermine science, its knowledge, or methodological claim to objectivity. It simply wishes to illustrate and explore further the possibility that, just as our understanding of nature can influence our understanding of ethics and politics, so too our ethical and political commitments may color our perception of the natural world.

ENDNOTES

1. Damien Veal, "Continental Philosophy and the Sciences," *Angelaki: Journal of the theoretical humanities* 10, no.1 (April 2005): 2.
2. Karen Cordrick Haely, *Objectivity in the Feminist Philosophy of Science* (London: Continuum, 2008), 6–12.
3. Haely, *Objectivity,* 12–15.
4. Thomas S. Kuhn, *The Structure of Scientific Revolutions* (Chicago: University of Chicago Press, 1962), 3–5, 43–51.
5. Kuhn, *Scientific Revolutions,* x, 111–34.
6. Haely, *Objectivity,* 15–23.

Index

Note: Entries with a page number, the letter n, and a second number indicate the reference is to a note on the designated page. For example, 131n4 indicates the reference is to note 4 on page 131.

Notes on Contributors

Douglas Al-Maini teaches philosophy at St. Francis Xavier University in Nova Scotia, Canada. Specializing in Ancient Philosophy with an emphasis on Plato, he has publications in the area from the Presocratics to the Stoics. He is currently working on a translation of the Roman Stoic Musonius Rufus.

Leah Bradshaw teaches political theory at Brock University in Ontario, Canada. Much of her work has grown out of considerations on the thought of Hannah Arendt. Recent publications have been on ancient and modern tyranny, configurations of love and women in the Western tradition, narrative and political identity, and the relation between emotions and reason in making political judgments.

Juhana Lemetti, Doc Soc Sci (Helsinki, 2006) is a postdoctoral researcher of the Academy of Finland at the University of Helsinki. He is the author of *Imagination and Diversity in the Philosophy of Hobbes* (Helsinki University Press, Helsinki, 2006) and a co-editor (with Eva Piirimäe) of *Human Nature as the Basis of Morality and Society in Early Modern Philosophy,* Acta Philosophica Fennica vol. 83 (Helsinki, 2007). Currently, Lemetti is writing a historical dictionary of the philosophy of Hobbes. He is also the associate editor of *Hobbes Studies.* His other areas of interest include political philosophy and the history of science.

Ingrid Makus teaches political theory at Brock University, including courses on liberal democracy, gender and political thought, and philosophy of law. Her publications include pieces on pity in Rousseau, Socratic reason in Nietzsche, and identity politics in Charles Taylor. Her recent work examines Simone de Beauvoir's contributions to a politics of responsibility.

Ahmed El-Sayed Abdel Meguid earned his Bachelor of Arts in Philosophy and Economics, Summa Cum Laude, at The American University in Cairo. He then earned his M.A. and is currently a Ph.D. candidate at the Philosophy Department, Emory University, Atlanta, Georgia. The title of his dissertation is "Humanity as historical-interpretative being: human nature and religious imagination between Kant and Ibn al-'Arab_.'" His research interests focus on German transcendental Philosophy and Phenomenology and Islamic Philosophy.

Dwayne Raymond earned a Ph.D. with distinction from the University of Western Ontario, Canada, in 2006. He is currently a visiting scholar at Texas A&M University, where he is writing a book on the origins of formal logic in ancient Greece. Dr. Raymond has given talks on the subject in Canada, America, Britain, France, and Finland.

Steven Robinson is currently the Acting Dean of Arts at Brandon University in Manitoba, Canada. He is an academic generalist. He began in physics and mathematics, and then made the shift into philosophy via the philosophy of biology and the evolution of ethics. Since discovering ethics proper he has concentrated on the history of value theory and metaphysics, and he wrote his dissertation on Plato's Symposium. He currently works on the origins of philosophy, and moral philosophy, within the social context created by ancient Greek democracy and tragic literature.

David Lewis Schaefer is Professor of Political Science at Holy Cross College in Worcester, Massachusetts, where he teaches courses on political philosophy and American political thought. A three-time fellow of the National Endowment for the Humanities, he is author of *The Political Philosophy of Montaigne* (1990) and *Illiberal Justice: John Rawls vs. the American Political Tradition* (2007), co-editor of Sir Henry Taylor's *The Statesman* (1992), and editor/ contributing author of *Freedom Over Servitude: Montaigne, La Boétie, and "Of Voluntary Servitude"* (1998).

Leonard Sorenson is Professor of Politics at Assumption College, Worcester, Massachusetts. He has published extensively on Jean Jacques Rousseau and James Madison. In addition, his most recent article, "Leo Strauss and the Defence of Western Civilization," has appeared in the journal, *The European Legacy.*

Paul Ulrich is Assistant Professor of Political Science, Philosophy, and Great Ideas at Carthage College in Kenosha, Wisconsin. He has published on Plato and has research interests in ancient political philosophy, early liberal thought, and Tocqueville.

Ann Ward is Associate Professor of Philosophy and Political Science at Campion College, University of Regina in Canada. She has published on Herodotus, Plato, Aristotle, Montesquieu and Kierkegaard. Her book is *Herodotus and the Philosophy of Empire* (2008). She has edited *Socrates: Reason or Unreason as the Foundation of European Identity* (2007), and co-edited with Lee Ward the *Ashgate Research Companion to Federalism* (2009). She is also an advisory editor for *The European Legacy.*

Lee Ward is an Associate Professor and Alpha Sigma Nu Distinguished Professor of Political Science at Campion College at the University of Regina. He is author of *The Politics of Liberty in England and Revolutionary America* (Cambridge, 2004) and co-edited with Ann Ward *The Ashgate Research Companion to Federalism* (Ashgate Publishing, 2009). He has published numerous articles on political thinkers including John Locke, Montesquieu, Aristotle, Plato, and Algernon Sidney.

Michael Weinman is currently on leave as Tutor at St. John's College, Annapolis and teaching ancient Greek philosophy at Ben Gurion University in Be'er Sheva, Israel. His book, *Pleasure in Aristotle's Ethics,* was released by Continuum Books in 2007. Weinman received the Alfred Schutz Oustanding Dissertation Award from the New School for Social Research in 2005. His articles on ethics and political philosophy have appeared in *International Studies in Philosophy* and *Essays in Philosophy.*

Mostafa Younesie is Assistant Professor of Greek/Persian political and lingual thought at the Humanities Faculty, University of Tarbiat Modares in Iran. He has published articles on Plato, Aristotle, Farabi, and Avicenna. His book is *Polis and Logos in Plato* (2008, in Persian), and he is currently writing his book *Qualities of the Text: Plato's Republic.* His current research is on the subject "Classical Greek Speculative Grammar."

Coleen Zoller is Associate Professor of Philosophy at Susquehanna University in Selinsgrove, Pennsylvania. She specializes in ancient philosophy and ethical theory. Her recent publications include "Plato on Hypothesis, Proportion, and the Education of Philosophers" in *Auslegung,* "The Pre-Critical Roots of Kant's Compatibilism" in *Philosophy and Theology,* and "To 'Graze Freely in the Pastures of Philosophy': The Political Motives and Pedagogical Methods of Socrates and the Sophists" in *Polis.*